REVISE BTEC
Application of Science UNIT 8
REVISION GUIDE

Series Consultant: Harry Smith

Author: Jennifer Stafford-Brown

THE REVISE BTEC SERIES

Application of Science Revision Guide	9781446902837
Application of Science Revision Workbook	9781446902844
Principles of Applied Science Revision Guide	9781446902776
Principles of Applied Science Revision Workbook	9781446902783

   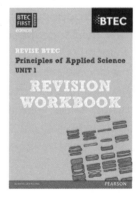

This Revision Guide is designed to complement your classroom and home learning, and to help prepare you for the external test. It does not include all the content and skills needed for the complete course. It is designed to work in combination with Edexcel's main BTEC Applied Science 2012 Series.

To find out more visit:
www.pearsonschools.co.uk/BTECsciencerevision

ALWAYS LEARNING

PEARSON

Contents

1-to-1 page match with the Revision Workbook ISBN 9781446902844

A small bit of small print

Edexcel publishes Sample Assessment Material and the Specification on its website. This is the official content and this book should be used in conjunction with it. The questions in *Now try this* have been written to help you practise every topic in the book. Remember: the real exam questions may not look like this.

Scientific equipment 1

You need to know the correct names for these pieces of scientific equipment, and know how each one is used.

Beaker

500
400
300
200
100

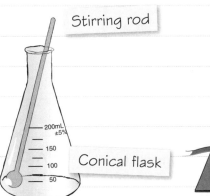

Stirring rod

Conical flask

200mL ±5%

150
100
50

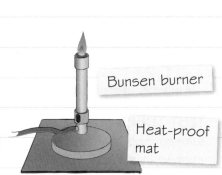

Bunsen burner

Heat-proof mat

👍 Used for mixing substances

👍 Lip for pouring

👍 Scale for measuring volume

👍 Wide base and narrow neck prevents spills

👍 Can be fitted with a bung

👍 Protects bench from heat

👍 Can be used with other hot equipment

Worked example

Label this diagram to show the name of each piece of equipment.　**(4 marks)**

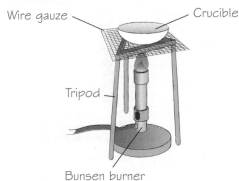

Wire gauze

Crucible

Tripod

Bunsen burner

You should also know what each piece of equipment is used for:
- Bunsen burner – for heating substances
- crucible – to hold solids or liquids being heated
- gauze and tripod – hold whatever is being heated in the right position above the flame.

Watch out! All this equipment gets hot.

Now try this

1　Match the pieces of equipment A to C shown in the diagram with the explanations 1 to 3, to explain why each is suitable.　**(3 marks)**

A　Stirring rod

B　Bunsen burner

C　Conical flask

1　Holds the liquid that is being heated.

2　Stirs the liquid without becoming hot.

3　Source of heat.

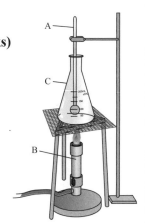

A

C

B

Scientific equipment 2

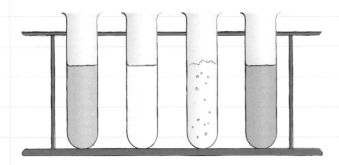

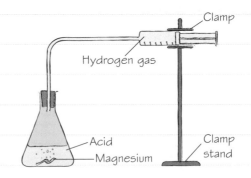

TEST TUBES hold small quantities of substances in a reaction. They must be stored in a TEST TUBE RACK to prevent the contents from spilling out.

A CLAMP and CLAMP STAND can hold a test tube or other container at a fixed distance from a heat source. Or they can support a piece of equipment.

Worked example

Sarah is carrying out an investigation. She needs to heat a small quantity of liquid in a Bunsen flame. Explain which piece of equipment would be best for this. **(3 marks)**

A boiling tube would be the best piece of equipment to use here because it is the right size to hold a small volume of liquid and it is safe to put in a Bunsen flame.

You need to be able to list the equipment needed and also give reasons for your choice. Think about what makes each piece of apparatus suitable.

Handle with care

A pair of TONGS is used to hold containers or substances that are being heated or tested.

Now try this

1   A student is carrying out an investigation to find out how much energy is released from different foods. Wearing safety goggles, they set light to the food and measure the temperature difference of some water in a test tube held above the flame.

Explain why it would be a good idea to use a clamp and clamp stand to hold the test tube full of water over the burning food. **(2 marks)**

Chemistry equipment

You need to be able to choose suitable chemistry equipment from that shown below and explain why you have chosen it.

You can measure pH with a pH meter, or universal indicator solution. You need this in neutralisation reactions to show when the neutralisation point is reached.

A piece of FILTER PAPER in a FILTER FUNNEL separates a solid from a liquid.

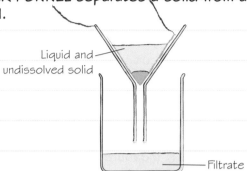

Liquid and undissolved solid

Filtrate

Measuring volume

Different sizes of MEASURING CYLINDER are available. To measure a smaller volume of liquid more PRECISELY, use a narrower cylinder.

A PIPETTE is used to measure or transfer small amounts of liquid.

Measuring temperature

- A THERMOMETER or temperature sensor measures temperature changes.
- A thermometer with 0.1°C intervals is more precise than one with 1°C intervals.
- Choose a thermometer with a suitable RANGE for your measurement.
- To measure temperatures below freezing, choose a thermometer that measures −10 to 100°C, not 0 to 100°C.

Measuring mass

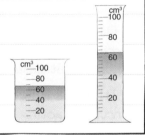

Cotton wool to stop acid 'spray' escaping

Marble chips

Dilute hydrochloric acid

Balance

DIGITAL BALANCES measure mass in grams or milligrams. A balance that measures to the nearest 0.01 g is more precise than a balance that measures to the nearest 0.1 g.

Worked example

To measure a small volume of liquid, which of the following pieces of equipment would be best to use?

A ☐ a beaker   B ☒ a measuring cylinder   C ☐ a test tube   D ☐ a funnel

A measuring cylinder is more precise than a beaker as it has more scale marks.

Now try this

1 Identify which piece of equipment you would use to:
   (a) transfer small amounts of liquid into a test tube **(1 mark)**
   (b) measure 10 ml of a substance **(1 mark)**
   (c) measure temperature changes **(1 mark)**
   (d) measure the pH of a solution. **(1 mark)**

Look at the equipment on pages 2–3 and decide which one would be most appropriate for each task.

Physics equipment

You need to be able to choose suitable physics equipment from that shown below and explain why you have chosen it.

Electrical equipment

A cell, battery or power pack provides the energy supply for a circuit. A power pack has an adjustable voltage output.

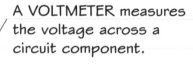

A VOLTMETER measures the voltage across a circuit component.

A MULTIMETER can be set to read current, voltage or resistance.

An AMMETER measures the current through a circuit component.

A RESISTOR can be connected in a circuit to reduce the current.

A filament bulb can be connected in a circuit to show that the circuit is working or to investigate its resistance.

A THERMISTOR can be used to control a circuit by acting as a temperature sensor.

A LIGHT-DEPENDENT RESISTOR (LDR) can be used to control a circuit by acting as a light sensor.

Choose an ammeter or voltmeter with a suitable range, such as 0 to 200 mA, or 0 to 6 V. A digital multimeter is useful because it has a selection of ranges.

A NEWTON METER measures weight or the force needed to pull an object.

A RULER or TAPE MEASURE is used to measure distance.

A STOPCLOCK or STOPWATCH measures time. LIGHT GATES can be set to measure time intervals, or speed.

Worked example

An engineer needs to calculate the acceleration of a model car on a ramp.

Describe what equipment he would need to calculate the car's acceleration. **(2 marks)**

Use light gates to measure the speed at two points and the time taken to go between the two points. Light gates provide better quality data than can be obtained using a stopclock to time the interval between the two points.

Now try this

1 Joanna is investigating the energy needed to push a trolley full of food round a supermarket.

Write a list of the equipment she would need to carry out the investigation and give reasons for your choices. **(3 marks)**

Only list the equipment that is relevant to the experiment. You may not gain full marks if you list apparatus that is not needed.

Biology equipment

You need to be able to choose suitable biology equipment and explain why you have chosen it.

A MICROSCOPE allows you to see samples and organisms that are too small for the human eye to see alone.

PETRI DISHES are used to grow bacteria safely in the laboratory.

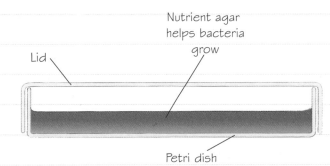

Nutrient agar helps bacteria grow

Lid

Petri dish

The sample or organism must be placed on a glass SLIDE. A glass COVER SLIP placed over the sample holds it in place and prevents the sample coming into contact with the microscope lens.

Worked example

A nurse needs to be able to work out the body mass index (BMI) of different people.

The formula for calculating BMI is:

BMI = mass in kg/(height in m)^2

List the equipment that will be needed for this measurement and give reasons for your choices. **(4 marks)**

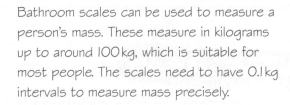

When measuring a person's height, take measurements to the nearest cm.

Bathroom scales can be used to measure a person's mass. These measure in kilograms up to around 100 kg, which is suitable for most people. The scales need to have 0.1 kg intervals to measure mass precisely.

Height can be measured using a tape measure up to at least 2 m with 1 cm markings. The person stands up straight against a wall with no shoes. A mark is made in line with the top of their head. The tape measure is used to measure their height up to the mark on the wall.

Now try this

1 Identify the piece of scientific equipment that can be used to:
   (a) see microscopic organisms   **(1 mark)**
   (b) place microscopic organisms on   **(1 mark)**
   (c) prevent a microscopic sample coming into contact with the lens   **(1 mark)**
   (d) grow a bacterial culture.   **(2 marks)**

2 A personal trainer is investigating the effects of exercise on the cardiovascular and respiratory systems. List all the equipment that she will need, and say why each piece is needed.   **(3 marks)**

Risks and management of risks

You need to know about the risks that may be harmful to health in an investigation and how to manage them.

A **RISK** is the harm that could be caused and the chances of it happening.

For example, if you hold a test tube over a Bunsen burner without using tongs there is a risk that you will burn your hand from either the flame or the heated test tube.

A **HAZARD** is something with the potential to cause harm, such as solutions that can irritate or burn the skin or eyes. Look at the hazard symbols on page 7.

Risk assessment

A **RISK ASSESSMENT**:

☑ identifies any possible hazardous substances or any activities that could cause harm

☑ considers the associated risks involved in using the substances or doing the activities

☑ evaluates the best way to manage and control the risks.

A **CONTROL MEASURE** is put in place to reduce the identified risks. For example, wearing protective clothing or using sterile techniques when handling microorganisms.

Worked example

Complete the missing pieces of information in the risk assessment below. **(3 marks)**

Hazard	Risk	Control measure
Bunsen burner	Setting light to hair	Tie long hair back
Hydrochloric acid	Can cause burns if in contact with the skin or eyes.	Use the lowest concentration possible and wear gloves and eye protection.

Think about the lab rules in your class and what you have to wear or do when you carry out an experiment – these are the control measures put in place to help prevent accidents from happening.

Now try this

1 Which **one** of the following is an example of a hazard? **(1 mark)**

  A ☐ a flammable chemical

  B ☐ a burnt arm

  C ☐ safety goggles

  D ☐ a beaker

Remember, a hazard is something with the potential to cause harm.

Hazardous substances and control measures

Many substances used in investigations could be harmful, so make sure you know what the hazard symbols on labels mean.

Corrosive

Avoid contact with the skin as it attacks and destroys living tissue.

For example, acids are corrosive so you should wear gloves and eye protection to protect your skin and eyes.

Flammable

These substances catch fire easily and should be stored in flame-resistant cupboards.

Keep flammable substances away from any flames, or test flammability in a fume cupboard.

Toxic

These substances can cause death if swallowed, breathed in or absorbed by skin.

To protect yourself, use this substance in a fume cupboard so that you do not breathe in harmful fumes. You should also wear eye protection and gloves to protect your skin.

Students are not allowed to use toxic substances. However, be aware of the risks of chlorine water, as small amounts of toxic chlorine vapour escape easily from the solution. Suitable control measures are to ventilate the room well and avoid inhaling the vapour.

Moderate hazard

This substance is not corrosive but will cause irritation to the skin, eyes or inside the body.

Gloves and goggles need to be worn when handling these substances.

Ultraviolet radiation

Ultraviolet (UV) radiation from the Sun can harm the skin and eyes. At low doses, it causes sunburn. Higher doses can lead to eye problems such as cataracts and blindness, and have the potential to cause skin cancer.

Worked example

Suggest **two** control measures you would take to minimise the risk of harm from UV radiation, when investigating how well different sunscreens block UV radiation from sunlight.   **(2 marks)**

There are two marks so you need to give **two** different control measures.

Sunglasses should be worn to protect the eyes from damage from UV light.

Suncream or clothing should be worn to cover the skin to protect the skin from UV light.

Now try this

1 Explain why a scientist should avoid contact with the skin when using hydrochloric acid.   **(2 marks)**

Protective clothing

Protective clothing can be worn to reduce the risks identified in a scientific investigation.

EYE PROTECTION
Is necessary to protect the eyes from a range of hazards including splashes from corrosive or hot liquids.

GLOVES
Are worn to protect the hands from harmful substances or to prevent the spread of infection if bacterial cultures are being grown. Gloves can also be worn to protect the hands from heat.

LAB COATS
If a lab coat came into contact with a Bunsen burner flame, it could be quickly removed, then extinguished.

Worked example

A technician is adding a dilute acid to a substance in a test tube to produce a reaction. He is wearing a lab coat, safety glasses and gloves.
Explain why the technician is wearing this protective clothing.   **(4 marks)**

He is wearing the safety glasses to stop the acid from splashing out at him and damaging his eyes. He has gloves on to protect his hands as the acid is corrosive and can cause skin burns. Lastly, he has a lab coat on to provide another layer of protection from splashes and spills from the acid.

Now try this

1 Explain why lab coats should be worn in a science lab.   **(2 marks)**

2 Explain when gloves should be worn in a science lab.   **(2 marks)**

3 Identify the protective clothing a technician should use if they are using a chemical that is an irritant.   **(2 marks)**

Handling microorganisms

There is a risk of infectious disease from handling some types of microorganism. You need to know the control measures for reducing this risk.

Risk	Control measure
Areas that could come into contact with food or drink and are potentially contaminated by bacteria.	☑ Bench surfaces should be wiped down with disinfectant after experiments with microorganisms.
Transfer of potentially harmful microorganisms from hands to mouth.	☑ Do not eat, drink, chew or lick sticky labels when working with microorganisms. ☑ Wash your hands with an antibacterial soap after handling any equipment.
Equipment used to handle microorganisms is not sterile.	☑ Sterilise all equipment after use. ☑ Follow instructions on how to dispose of Petri dishes once you have finished with them.

Growing microorganisms

Microorganisms may be cultured (grown) in a Petri dish. The microorganisms are safe, but it is easy to also grow dangerous microorganisms from the air or from hands. There are easy ways to reduce the risk.

- Tape the lid to the dish to stop microorganisms from the air getting in.
- Incubate the Petri dish at 25°C. Temperatures higher than this encourage the rapid growth of bacteria and that might include pathogens.

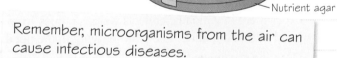

Petri dish — Lid — Nutrient agar

Remember, microorganisms from the air can cause infectious diseases.

A microbiologist handles microorganisms. Describe **one** control measure that can help to prevent the transfer of these microorganisms from hands to mouth. **(2 marks)**

They should not eat in the lab as they may have microorganisms on their hands. If they then picked up food with their hands, the microorganisms would be transferred to the food which would then be put in the person's mouth to eat. So by not eating in the lab it helps to prevent this risk from happening.

1 Imad is growing cultures in Petri dishes to investigate the effect of different hand washes on the growth of bacteria.

   Give **two** precautions he would need to take when carrying out this experiment to minimise the risk, and explain why each precaution is necessary. **(4 marks)**

Microorganisms and wildlife

Some investigations might need you to handle wildlife – usually small invertebrates. There are control measures which need to be put in place for this sort of experiment.

Wildlife

Wild animals may have been in contact with microorganisms. When handling wild animals wash your hands with antibacterial soap, wipe down surfaces following the experiment and clean equipment appropriately.

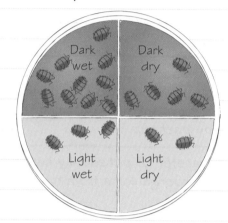

> Wild animals need to be treated with respect and returned to their own environment after use.

Disinfectant kills microorganisms and can be used on surfaces and equipment.

An AUTOCLAVE heats equipment to high temperatures to kill microorganisms and is a more effective method than using disinfectant.

Worked example

An environmental scientist is carrying out an experiment in which he collects water samples from a pond to find out about the wildlife living in the pond.

Describe the control measures that should be used to carry out this experiment.   **(2 marks)**

He should make sure that he does not eat or drink anything when he is collecting or analysing the samples as he may transfer microorganisms into his mouth from the pond. He should also wash his hands with an antibacterial soap after collecting the samples and again after handling the samples.

← Remember, soil could also contain microorganisms that can have a risk of infectious disease.

Now try this

1 Describe why it is important to wash hands with an antibacterial soap after having handled soil samples.   **(2 marks)**

2 A dentist will use an autoclave to clean some of their equipment.

Explain why an autoclave is used rather than disinfectant spray.   **(2 marks)**

Dependent and independent variables

You need to be able to identify the key VARIABLES in an investigation.

The DEPENDENT VARIABLE is the outcome that is measured during the investigation. What happens to it DEPENDS on the changes you deliberately make to the INDEPENDENT VARIABLE.

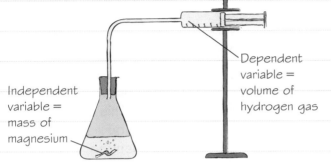

Independent variable = mass of magnesium

Dependent variable = volume of hydrogen gas

It is not possible to have a dependent variable without an independent variable – if you change one variable, you look for its effect on another.

Investigation into the volume of hydrogen gas produced when the mass of magnesium is changed.

QUANTITATIVE variables have values that can be measured, such as weight, temperature and mass. Some variables are difficult to measure and you might have to describe them QUALITATIVELY, such as the brightness of a flame.

Worked example

In the following investigation, what is the dependent variable and what is the independent variable? Give reasons for your answer. **(4 marks)**

An experiment was conducted to find out how exercise participation affects body mass index. 3 groups of 20 males aged between 18 and 21 years old took part in the experiment. Group A took part in no exercise, Group B took part in some exercise and Group C took part in a lot of exercise.

The dependent variable is the body mass index as this is the value that may change as a result of taking part in the differing amounts of exercise. The independent variable is the amount of exercise each group participated in.

Now try this

1   A scientist is investigating the effect of smoking on vital lung capacity. They measure the vital lung capacities of 3 groups of people, Group A have never smoked, Group B smoke 10 cigarettes per day and Group C smoke 20 or more cigarettes per day.

The dependent variable is the outcome that is measured during the investigation.

In this investigation, state which one of these variables is the:
(a) dependent variable   (b) independent variable.   **(2 marks)**

2   A gardener is trying to increase the height of sunflowers and wants to find out the optimal quantity of fertiliser that she should use in order to do grow taller sunflowers. She adds different quantities of fertiliser to 10 different plants and records the height of each plant when it is finished growing.

State which one of these variables is the:
(a) dependent variable   (b) independent variable.   **(2 marks)**

Control variables

A CONTROL VARIABLE is a quantity that remains constant. Most experiments will have one or more control variables.

It is important to control variables so that the investigation is a fair test of the hypothesis, and the results are valid.

In an experiment to find out how the mass of salt added to water affects the boiling temperature of water, the AIR TEMPERATURE and the AIR PRESSURE are both the same – these are the control variables.

Science experiment	Independent variable	Dependent variable	Control variables
How is the rebound height of a squash ball affected by different coloured dots marked on the balls?	The dot on the squash ball – this indicates the speed of the ball.	The rebound height – this is what you are measuring.	Same drop height. Same person dropping the squash ball so the method of release is the same each time.

Worked example

In the following investigation, identify the control variables.

(2 marks)

A scientist is trying to find out how well different antacid tablets neutralise acid. They use $5\,cm^3$ $0.1\,M$ HCl for each test and add small quantities of crushed antacid tablet until the acid is neutralised.

The control variables are the ones that are kept constant for each experiment. Therefore, the concentration and volume of the acid solution are control variables, along with the quantity and size of the antacid used.

Independent variable – what I will change.

Dependent variable – what I will measure or observe.

Control variable – what I will keep the same.

You need to be able to identify some of the appropriate variables to control, and also describe how to control these variables.

Now try this

1. In the worked example above, describe how you would control the volume of hydrochloric acid.

   (2 marks)

2. An experiment was carried out to find out how the mass of an object affected its speed at the bottom of a 2 metre long ramp.

   Identify **three** variables that have to be controlled in this experiment.

   (3 marks)

Measurements

You need to be able to identify what sorts of measurements to take in an investigation, and the units to use for each measurement.

Metric units

You need to state the UNITS with any measurement.

The line AB is 35 mm long. You can also write this as 3.5 cm.

Appropriate units

You need to be able to choose sensible units.

Litres   Millilitres

Most of the units of measurement used in the UK are METRIC units.

You can convert between metric units by multiplying or dividing by 10, 100 or 1000.

Length	**Weight**	**Volume or capacity**

Length

÷1000 ⟲ km ⟳ ×1000
÷100 ⟲ m ⟳ ×100
÷10 ⟲ cm ⟳ ×10
mm

Weight

÷1000 ⟲ tonne ⟳ ×1000
÷1000 ⟲ kg ⟳ ×1000
÷1000 ⟲ g ⟳ ×1000
mg

Volume or capacity

÷100 ⟲ litre ⟳ ×100
÷1000 ⟲ cl ⟳ ×1000
÷10 ⟲ ml or cm³ ⟳ ×10

Worked example

MATHS SKILL

A biologist is using a microscope to view bacteria in a Petri dish. The magnification he is using is in micrometres (μm).

(a) Identify the correct standard form for measurements in micrometres.   **(1 mark)**

   A ☐ 10^{-3} **B** ☒ 10^{-6} **C** ☐ 10^{-5} **D** ☐ 10^{-2}

(b) State the metric measurements that are given at 10^{-9}.

   nanometres.

Micrometres are $\dfrac{1}{1\,000\,000}$ which is 10^{-6}.

A nanometre is measured at $\dfrac{1}{1\,000\,000\,000}$ which is 10^{-9}.

Now try this

1 How many mm are in:
   (a) 9.1 cm   **(1 mark)**
   (b) 0.7 cm   **(1 mark)**

2 Ultraviolet waves used in fluorescent lamps have a wavelength of 10 nm. What is this number in metres?   **(1 mark)**

Units of measurement

You should always include the correct units of measurement when you record data or give the answer to a calculation.

Type of measurement	Unit	Symbol
Length	metre	m
	centimetre	cm
	millimetre	mm
	kilometre	km
Mass	kilogram	kg
	gram	g
Time	minute	min
	second	s
Temperature	Celsius	°C
Area	square metre	m^2
Volume	cubic metre	m^3
	litre	l
	millilitre	ml
Density	grams per cubic centimetre	g/cm^3
Speed or velocity	metres per second	m/s
Acceleration	metres per second squared	m/s^2
Frequency	hertz	Hz
Force	newton	N
Power	watt	W
	kilowatt	kW
Energy, work	joule	J
	kilojoule	kJ
Potential difference	volt	V
Resistance	ohm	Ω
Current	ampere	A

You need to know the units and the symbol for each type of measurement.

- -

Now try this

1 Which of the following is the correct unit for acceleration? **(1 mark)**

   A ☐ m/s   **B** ☐ m/s^2 **C** ☐ ml   **D** ☐ W

2 Give the correct symbol for a unit of power. **(1 mark)**

3 Identify the units that you would use to measure the frequency of a wave. **(1 mark)**

4 Identify the type of measurement that is measured in watts. **(1 mark)**

Accurate and precise measurements 1

ACCURACY and PRECISION have different meanings in science. All measurements need to be accurate and precise in order to give meaningful conclusions.

Accuracy

If your measurements are close to the correct value they are ACCURATE.

This archer is accurate – the arrows are all close to the centre.

Precision

If you measure the same thing lots of times and your measurements are close together then they are PRECISE.

This archer is precise – the arrows are all close to the same spot.

Precise but not accurate

It is possible for a measurement to be precise but not accurate. If you weigh the same thing 10 times and get measurements that are very similar or the same then your measurements will be PRECISE.

These scales are not set correctly to 0 so your measurements won't tell you the real weight of the object, so they won't be ACCURATE.

Worked example

MATHS SKILL

A student weighs 4 samples of powdered metal labelled A, B, C and D. The actual mass of each sample is 8.14 g. The student's results are shown below.

Sample	Mass (g)
A	8.35
B	8.33
C	8.33
D	8.35

(a) Calculate the mean mass for the powdered metal from these results. **(1 mark)**

(8.35 g + 8.33 g + 8.33 g + 8.35 g) ÷ 4 = 8.34 g

(b) Explain if these data are accurate. **(2 marks)**

These data are not accurate as the true mass of the powdered metal is 8.14 g, which is 0.2 g different from the student's measured mean mass.

(c) Explain if the data are precise. **(2 marks)**

Yes, the data are precise as the results are all within 0–0.02 g of each other.

Precision can be worked out by seeing how each measurement differs from the mean value – in this case, two values are 0.01 g higher than the mean value and two values are only 0.01 g lower than the mean value. So it can be said that these results are precise.

Now try this

1 Describe what is meant by accurate measurements. **(1 mark)**

2 Jane carries out the same test 3 times on the same person to work out the power of their legs.

The results are shown below:

20 W   25 W   30 W

Describe if these results are precise. **(2 marks)**

Accurate and precise measurements 2

You need to read scales on measuring instruments carefully.

You should always read scales correct to the nearest SMALL subdivision.

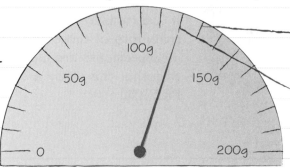

On this scale there are five divisions between 100 g and 150 g, so each division represents 10 g.

The scale reads 120 g.

Consistent measurements

When you carry out an experiment it is important to always measure in the same way. When you read the scale on a MEASURING CYLINDER you should always read to the bottom of the MENISCUS.

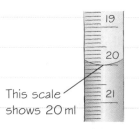

This scale shows 20 ml

Worked example

In an experiment to measure the heat energy produced by burning different foods, John decided to measure the temperature rise of 100 cm³ water. He used a beaker to measure the volume of water.

(a) Give **two** ways he could improve this measurement. **(2 marks)**

He should use a measuring cylinder rather than a beaker as it is more accurate. He should always read to the bottom of the meniscus in order to get an accurate measurement.

(b) Explain why it is important that John uses the same volume of water each time. **(2 marks)**

The same volume of water should be used as it is the control variable. If the volume differs between experiments it could affect the calculation of energy produced by each food, since larger volumes of water need more energy to increase its temperature by 1 degree.

Now try this

1 Explain why, if a person took readings from the top of the meniscus in a measuring cylinder, the results would not be accurate. **(2 marks)**

Range and number of measurements

You need to choose a suitable range and number of measurements to be made in an investigation, and to explain why you chose them.

How many measurements?

A trial run of your experiment will help you to plan:

- the RANGE (maximum and minimum) of the independent variable
- the intervals for your independent variable
- how many measurements to take.

Remember that a bigger range of results will help find a pattern or trend.

It is best to choose values that are evenly spaced, if possible, and to use at least six different values.

Repeating measurements

It is a good idea to repeat measurements at least three times to check that your results are REPEATABLE.

- If the values vary slightly a MEAN can be taken – the mean of several measurements is more accurate than a single value.
- If one of the measurements is very different (ANOMALOUS) this is easy to identify (see page 24 to remind yourself about anomalies).

Worked example

Ann carries out a trial run to find out how temperature affects the rate of a reaction.

Temperature (°C)	Rate of reaction (g)
10	15
30	100

Suggest how many measurements Ann should take in this range. Give a reason for your answer.   **(2 marks)**

5 °C intervals could be used as there is a very large increase in the rate of reaction between 10 °C and 30 °C. Therefore, she should take five measurements. By including more temperature measurements it will help to show the relationship between the temperature and rate of reaction.

Now try this

1 Lauren wrote this plan.

I am investigating the relationship between the extension of a spring and the weight attached to it, so I need to hang a weight on a spring and measure the increase in length. I will then do the experiment again with different weights.

Explain **two** ways Lauren's plan could be improved.

(4 marks)

For each improvement you need to describe a feature for the first mark and say **why** for the second mark.

Writing a method

When you write a method for an investigation you should provide enough detail so that another person can use this information to carry out exactly the same investigation as you.

Method

✓ List of equipment.

✓ Quantity of substances needed.

✓ Description of any special techniques (and why they're important).

✓ How many measurements to take (and how to know if any need repeating).

✓ Labelled diagram (with reasons for choosing each piece of equipment).

Thermometer
Stirrer
Clamp
Calorimeter
Water
Spirit lamp
Fuel

Apparatus to investigate the heat energy produced by burning different fuels

---

Worked example

Steve was investigating the effect of temperature on the colour of thermochromatic materials used in novelty mugs, which reveal a picture when a hot drink is poured in. Part of his method is shown below.

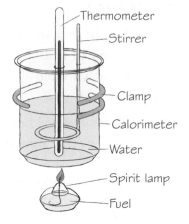

Use water at the following temperatures:

50 °C, 60 °C, 70 °C, 80 °C, 90 °C and 100 °C

Pour water straight into the mug and see if the image is revealed. Return the mug to room temperature between each test.

He found that the image was revealed at 90 °C. Describe how this method could be improved. **(2 marks)**

The image has been found to be revealed at 90°C but the temperature that was required to make this change could have been anything from 81°C through to 90°C, as the temperature intervals used are quite far apart. It would be better to use smaller intervals of temperature to find the exact temperature that revealed the image.

Now try this

1 Amir has written a method to test his hypothesis that the heavier an object, the faster it will travel down a ramp. His method is shown below.

- Place electric timer gates at the top of the ramp nead at the bottom of the ramp.
- Place the trolley at the top of the ramp.
- Push the trolley down the ramp.
- Take a note of the times recorded on the electric timer gates.

Explain if this method tests the hypothesis 'the heavier an object, the faster it will travel down a ramp'.　　**(3 marks)**

To gain full marks for an investigation, your method must explain how it will test the hypothesis.

Hypotheses

A HYPOTHESIS states clearly what you expect to happen in an investigation, based on relevant scientific ideas.

Sue's company is investigating which training shoe has better 'grip' on tarmac.

Hypothesis: the greater the area of tread, the greater the force required to move the shoe. (Different tread patterns is the independent variable; force is the dependent variable.)

 A hypothesis relates the independent variable to the dependent variable.

This is because a bigger tread means there will be more contact area for friction between the shoe and tarmac.

 Explain what you think will happen in the investigation and why you think this will happen.

Test the hypothesis by changing the area of shoe in contact with the ground and measuring the force required to move the shoe along a section of tarmac.

 It is best to write a hypothesis that you can test easily with an investigation.

Worked example

Simon is carrying out an investigation to find out how the dose of penicillin used affects the growth of bacteria.

He decides to place discs of filter paper containing different masses of penicillin on different Petri dishes containing a bacterial culture in agar. He will then see where there are the largest zones of inhibition of bacterial growth to work out which dose of penicillin is the most effective in stopping the growth of bacteria.

State an appropriate hypothesis for this experiment and give a scientific reason. **(2 marks)**

If the dose in milligrams of penicillin is increased, the diameter of the clear area around the penicillin disc will be increased. This is because antibiotics kill or control the growth of microorganisms, and a higher dose of penicillin means more bacteria will be killed.

Zone of inhibition

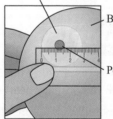
Bacterial growth

Penicillin disc

 The mass of penicillin is the independent variable as this is determined by the scientist.

The hypothesis needs to state what is predicted to happen to the bacteria in response to the increased doses of penicillin.

Now try this

1   Charles writes the hypothesis:

The lower the concentration of penicillin, the greater the reduction in growth of bacteria.

Explain if this hypothesis is written correctly. **(2 marks)**

 Your experiment might actually disprove your hypothesis! This is fine, as it still tells you important information about the variables in your investigation.

Qualitative and quantitative hypotheses

Qualitative hypothesis

A QUALITATIVE hypothesis describes what is expected to happen but does not include a numerical prediction. To say that something will increase or decrease would be appropriate for a qualitative hypothesis.

Quantitative hypothesis

A QUANTITATIVE hypothesis gives an expected value, percentage or factor change. For example, the current through the resistor will double when the voltage across it is doubled.

Worked example

A dietician is carrying out an investigation to test the idea that eating more vegetables decreases the risk of developing heart disease.

State if this is a qualitative or a quantitative hypothesis. **(1 mark)**

This is a qualitative hypothesis.

This hypothesis is qualitative as it describes what she expects to find without giving a value for how much the risk will be reduced.

Worked example

A health worker is researching the effect of exercise on reducing the risk of coronary heart disease (CHD).

They state the hypothesis: taking part in regular physical activity will reduce the risk of CHD by 50%.

State if this is a qualitative or a quantitative hypothesis. **(1 mark)**

This is a quantitative hypothesis.

An expected value of 50% has been stated which means the hypothesis is quantitative.

Now try this

1 State if the following hypotheses are quantitative or qualitative.
   (a) For every 5 °C increase in temperature of water there will be a 2 g increase in the quantity of sugar dissolved. **(1 mark)**
   (b) The taller a person is, the larger their vital lung capacity. **(1 mark)**

2 The hypothesis below is a quantitative hypothesis.

   For every 10 g of fertiliser added to the soil there will be a 5 per cent increase in the height of the sunflower.

   Change this hypothesis to a qualitative hypothesis. **(1 mark)**

Planning an investigation

It is important to write a plan before you start, so you can make sure the experiment will help test your hypothesis. If another scientist repeats your experiment they will be able to carry it out in exactly the same way.

Your plan needs to describe each of these stages in a scientific investigation.

Think of an idea and then ask a question about it	State your hypothesis and make sure it can be tested by an experiment.
Carry out background research	What scientific ideas could explain how or why this happens?
Make a prediction about what you think will happen	
Write your method, as steps in the correct order	Include: • the equipment, and reasons for choosing it • which variable will be changed, and why • which variables will be controlled • the number and range of measurements to take • how any risks can be reduced
Carry out an experiment to test the hypothesis	
Analyse results	Do they agree with your prediction?
Draw a conclusion	Explain what has happened in the experiment and why you think it has happened.

In the exam, you could also be asked to apply these skills to planning an investigation in an unfamiliar situation.

Now try this

1 Plan an investigation to find out about the effect of wire length on resistance in a circuit. **(4 marks)**

Learning aim A sample questions

This page looks at the type of questions you might be asked to answer as part of the Unit 8 Scientific Skills paper. These questions cover material from Learning aim A. The experiment investigates how temperature affects the rebound height of a squash ball. The experiment explores gravitational potential energy, kinetic energy and the effect of heat on the transfer of energy.

Worked example

Hypothesis
Write a hypothesis for how the temperature of the squash ball affects the height of the rebound.　**(2 marks)**

The hotter the temperature of a squash ball, the higher the rebound height.

> This hypothesis is good but not complete. It would be better if it contained a quantitative prediction. For example, doubling the temperature of the squash ball will double the rebound height.

Worked example

Plan
Write a plan for this investigation.　**(6 marks)**

Equipment
Large pieces of paper, pencil, sticky tape, chair, measuring tape/metre rule, 1 squash ball, water baths, thermometer.

> This is a good list of equipment because it contains everything you might need and all the equipment listed is relevant to the investigation.

Method
1. Use a tape measure to measure out a height of 2 m and width of 1m on a wall.
2. Place sheets of plain paper over this area on the wall and attach with tape.
3. Using the tape measure, mark each cm on the paper. Draw horizontal lines across the whole width of the paper.
4. Heat a squash ball in a water bath at 10°C for 10 minutes.
5. Stand on a chair and hold the ball at the 1.5 m mark on the paper.
6. Holding out a straight arm, release the ball.
7. Another person observes the ball drop and marks the rebound height taken from the bottom of the ball onto the wall paper.
8. Record the rebound height on a table.
9. Repeat steps 4–8 of the experiment with the squash ball heated at the following temperatures: 10°C, 15°C, 20°C, 25°C, 30°C, 35°C, 40°C.

> This is an excellent method because it is clearly set out and it clearly tests the hypothesis.

> It is clear that the squash ball will be heated in the same way each time. This is the independent variable. Other details such as the height the ball is dropped from are also kept the same.

> This is a good range of suggested measurements.

Now try this

1　Plan an experiment to test how the quantity of salt in 100 cm³ of water affects the boiling point.　**(6 marks)**

> Remember to include the equipment that you will need to use and a method to show how the experiment will be carried out.

Tables of data

Recording data in a TABLE helps to organise the information and makes it easier for the reader to understand what the data are showing.

Remember to:

- ☑ arrange data in columns
- ☑ put the independent variable in the first column
- ☑ give each column a heading that explains the values that it contains
- ☑ include units in the column heading where appropriate – not all values have units
- ☑ arrange data in order of the independent variable, usually in increasing order.

The table below shows the results of an investigation into the relationship between force, mass and acceleration for a trolley travelling down a ramp. The mass was kept constant (control variable).

INDEPENDENT VARIABLE

shown in increasing order (force is increasing).

Force (N)	Acceleration (m/s²)
1.0	0.24
2.0	0.50
3.0	0.73

DEPENDENT VARIABLE

Units are included under each heading.

Worked example

John is carrying out an investigation to find out how different temperatures affect the mass of sugar that can be dissolved in water.

Draw a table with suitable headings for John to record his results in.

(2 marks)

Temperature (°C)	Mass of sugar (g)

The independent variable (the one John changes) is temperature so this goes in the first column.

The dependent variable is the one that changes as a result of changing the temperature, so the mass of sugar goes in the next column.

Units for each measurement are included in the table headings.

Now try this

1 Describe the pattern that you can see in the table below. **(1 mark)**

Temperature (°C)	Mass of sugar (g)
10	5
15	8
20	11
25	14

A well-designed table will help you see any pattern that exists in the data – this will help you to start to draw conclusions about what is happening in your investigation.

Identifying anomalous results from tables

An ANOMALOUS RESULT (also called an ANOMALY) is a measurement that falls outside the range of the other results. It does not fit the pattern of the other results.

An anomalous result usually means that you have not taken the measurement accurately.

It is usually easier to spot an anomalous result when data are written in a table.

Worked example

Saima carried out an experiment to find out how a spring extended when different masses were attached to it. The results are shown in the table below. Explain which result is anomalous.

(3 marks)

Mass (g)	Extension (cm)
100	2.4
200	4.0
300	5.3
400	6.9
500	7.5
600	10.1
700	11.5

The reading of 7.5 cm at 500 g mass is anomalous as it does not fit the pattern of the other results.

The extension goes up by about 1.6 cm for every 100 g increase in mass. However, the extension at 500 g is only a 0.6 cm increase so it is not following the same pattern as the other results.

Putting the data in increasing order of the independent variable helps show up the anomalous result. Anomalous results can be even easier to spot on a graph.

You should also be able to suggest reasons why the reading is anomalous.

Now try this

1   An experiment was carried out to find out how the mass of salt added to water affected the boiling point.

The results are shown in the table opposite.

(a) Which result is anomalous?   **(1 mark)**
(b) Explain your answer to part (a).   **(2 marks)**

Mass of salt (g)	Boiling point (°C)
5	99
10	95
15	94
20	85
25	79
30	75

Identifying anomalous results from graphs

An anomalous result is sometime easier to spot on a graph rather than in a table.

This graph shows the results of the investigation in the extension of a spring from the previous page. Plotting the graph shows the anomaly clearly.

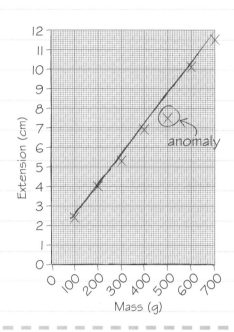

Anomalous results usually occur from human error. This means the person carrying out the experiment has made at least one mistake!

Worked example

James carried out an experiment to find out how quickly the boys in his class could run 100 m. The results are shown below.

Runner	Time (s)
1	15.5
2	12.6
3	11.2
4	13.4
5	14.8
6	30.2
7	14.6
8	12.7

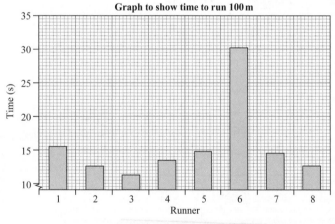

Which of these results is an anomalous result?   **(1 mark)**

The result for runner 6 (30.2 seconds) is anomalous.

The time for runner 6 is approximately twice as much as the other runners' times – it is anomalous because it is outside the range of the other results.

Now try this

1   Give **one** reason why the 100 m run time for runner 6 could have been so much slower than all of the other runners.

(1 mark)

Excluding anomalous results

Anomalous results can have a significant effect on any calculations from your results. Handle them appropriately so that they do not affect the conclusions you make.

If you spot an anomalous result but can't explain it, repeat that measurement. If the repeated measurement is not anomalous, use that and discard the first result as it must be due to an error in the way you carried out the experiment.

In an experiment to investigate the effect of the mass of salt on the boiling temperature of water, you could repeat a measurement for an anomalous result, making sure that:

- the mass of salt is measured correctly (check that the balance is set to zero before adding the salt)
- the same volume of water is used as in the other tests
- the water is boiling when you measure the temperature.

Excluding anomalous results

If you can explain why you have an anomalous result then it should be ignored and left out of any calculations, such as calculating the mean of the results.

If you identify an anomalous point on a graph, leave it on the graph but ignore it when drawing a best-fit line.

Worked example

James carried out an experiment to find out how quickly the boys in his class could run 100 m. The results are shown below. Calculate the mean time it took the boys to run 100 m.   **(2 marks)**

Runner	Time (s)
1	15.5
2	12.6
3	11.2
4	13.4
5	14.8
6	30.2
7	14.6
8	12.7

The result for runner 6 is anomalous so I will leave it out of the calculation.

$15.5 + 12.6 + 11.2 + 13.4 + 14.8 + 14.6 + 12.7 = \dfrac{94.8}{7}$

$= 13.5 \text{ seconds}$

As the times are all given to 1 decimal place, the mean value should be rounded up or down to 1 decimal place.

Now try this

1 Sol investigates the power of a person's legs by getting them to carry out a vertical jump test.

Each person carries out the test 4 times. The results are shown below.

(a) Describe any anomalous data in this investigation.   **(1 mark)**

(b) Complete the table with the mean jump height for each person.   **(4 marks)**

Person	Jump height (cm)	Mean jump height (cm)
A	20.2, 22.4, 21.2	
B	15.6, 16.8, 17.9	
C	12.2, 22.5, 21.9	
D	18.9, 19.8, 17.9	

Calculations from tabulated data

Repeating your experiment several times helps to make your results more reliable. The repeat results can then be used to calculate a mean value. Anomalous results should not be included in the calculation of a mean.

Worked example

MATHS SKILL

A physiologist carried out an investigation into the effect of exercise on heart rate.
She worked with a group who each took their heart rate while resting and then increased the amount of exercise they did each time. Here are the results:

Length of exercise (min)	Heart rate (beats per minute)				
	1st volunteer	2nd volunteer	3rd volunteer	4th volunteer	Mean
0	80	80	72	72	76
2	132	136	124	120	128
4	156	120	152	156	146

(a) Calculate the mean heart rate for each different period of exercise. **(3 marks)**

76 bpm, 128 bpm, 146 bpm

(b) She realised that one of the results was anomalous. This was the heart rate for the second volunteer at 4 minutes. She decided to re-calculate the mean heart rate for 4 minutes of exercise but leave out the anomalous result. Calculate the mean heart rate after 4 minutes of exercise without the anomalous reading. **(2 marks)**

156 + 152 + 156 = 464

$$\frac{464}{3} = 155 \text{ bpm}$$

To calculate the mean you need to add up the heart rate for each volunteer and then divide by the number of volunteers:

0 minutes = 80 + 80 + 72 + 72
= 304

$$\frac{304}{4} = 76 \text{ bpm}$$

Now try this

1 Brody carries out an investigation into the effect on temperature of changing the concentration of hydrochloric acid added to sodium hydroxide.

He carried out 3 tests for each of the different concentrations of HCl.

Concentration of HCl (M)	Temperature change (°C)
1	5, 4, 4
2	7, 6, 6
3	8, 8, 7
4	10, 12, 11
5	14, 13, 18

Calculate the mean temperature change for each concentration of HCl, making sure any anomalous data are excluded from the calculations. **(5 marks)**

Calculations from tabulated data – using equations

It is useful to include space in your table for the results of any calculations.

Worked example

Shannon investigated how the acceleration of a trolley on a slightly sloping ramp depends on the force applied to the trolley. Complete the table to find the acceleration for each applied force. You will need to use the equation:

$$\text{acceleration (m/s}^2) = \frac{\text{change in velocity (m/s)}}{\text{time taken (s)}}$$

For the 0.2 N force the calculation is:

$$\text{acceleration} = \frac{(\text{second velocity} - \text{first velocity})}{\text{time taken}}$$

$$= \frac{(2.11 - 0.1)}{1.59}$$

(3 marks)

Force (N)	First velocity (m/s)	Second velocity (m/s)	Time taken (s)	Acceleration (m/s²)
0.2	0.10	2.11	1.59	1.26
0.4	0.12	3.23	1.29	2.41
0.6	0.21	3.61	0.89	3.82

Worked example

Sam burned 6 different crisps to find out how much energy they transferred to 20 g of water. Calculate the energy transferred when the Ridges were burned. **(2 marks)**

Use the following equation: $q = m \times C \times \Delta T$

q = energy transferred (J)
m = mass (g)
C = specific heat capacity of water = 4.2 J/°C/g
ΔT = change in temperature (°C)

Crisp	Initial temperature (°C)	Final temperature (°C)	Temperature increase (°C)
Ridges	19	29	10

Ridges: $q = 20 \times 4.2 \times 10 = 840 J$

Now try this

1 Charlotte is investigating the energy change when zinc reacts with copper sulfate solution.

   (a) Calculate the temperature increase during the experiment. **(2 marks)**
   (b) Charlotte used 50 cm³ of copper sulfate. Use the equation in the second worked example to calculate the heat given out in the experiment. **(2 marks)**

	Experiment
Initial temperature (°C)	17.5
Final temperature (°C)	27.9
Temperature increase (°C)	

Significant figures

SIGNIFICANT FIGURES are the digits included in a measurement or a calculation that show how accurate the value is.

Digital bathroom scales measure body mass in kg and show at least one decimal place:

45.07 kg

most significant   least significant

the 0 is significant, because it is between other numbers and its value is important

You always start counting significant figures from the left.

125 cm has three significant figures.

27.05 g has four significant figures.

Rounding up or down

125 cm rounded to two significant figures is 130 cm

because the digit after the 2 is 5, so you round up.

27.04 g rounded to three significant figures is 27.0 g

because the digit after the 0 is 4, so you round down.

0.127 g rounded to two significant figures is 0.13 g

because the digit after the 2 is 7, so you round up.

Be careful to round numbers up or down – not just chop off the last few digits.

Worked example

State the values below to 3 significant figures:

(a) 80945826   (1 mark)
(b) 0.000684597   (1 mark)

(a) 80900000   (b) 0.000685

Do not write 4.6875 or even 4.69, because you have not measured either distance or time to this accuracy.

Worked example

A person runs 30 m in 6.4 seconds. Calculate their speed.   (2 marks)

These measurements are both given to two significant figures so the speed should be given to two significant figures.

speed (m/s) = distance (m)/time (s)

$$= \frac{30\,m}{6.4\,s}$$

$$= 4.7\,m/s$$

Now try this

Remember to always include the units of a measurement when you give the answer to a calculation.

$R = \dfrac{V}{I}$

1 Calculate the resistance of a circuit with a voltage of 5.0 V and a current of 3.0 A.   (1 mark)

2 Calculate the voltage of a circuit that has a resistance of 6.0 Ω and a current of 2.0 A.   (1 mark)

Bar charts 1

In a BAR CHART the heights of the columns are the values of the dependent variable. These are plotted against the independent variable. This is a good way to compare values.

When you draw a bar chart or a line graph:

This is the y-axis. ➡

the dependent variable is always plotted on the vertical axis

The dependent variable is the value that you are measuring in your experiment.

Dependent variable

0

Independent variable

the independent variable is plotted on the horizontal axis

This is the x-axis. ➡

The independent variable is the value that you are changing in your experiment.

When to use a bar chart

Bar charts are usually used when:

- the independent variable is qualitative, e.g. eye colour, flavour of crisps
- the dependent variable is quantitative, e.g. time for a reaction to happen, length of a spring.

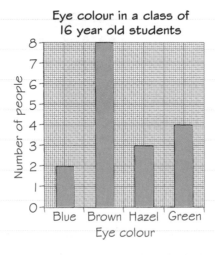

Eye colour in a class of 16 year old students

Now try this

1 Identify if a line graph or bar chart would be the most suitable method of presenting the following data from investigations:

(a) The energy produced from burning different foods. **(1 mark)**

(b) The effect of increasing mass on the speed of a car. **(1 mark)**

(c) The effect of increasing drop height of a paper helicopter on the time taken to fall. **(1 mark)**

Bar charts 2

Drawing bar charts

Remember to include all of the following in a bar chart:

✓ A title that provides a summary of the chart content.

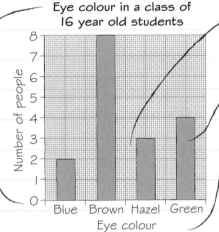

Eye colour in a class of 16 year old students

✓ Leave a gap between each bar to make the chart easier to read.

✓ Bars with equal widths.

✓ The scale on the y-axis always starts at zero.

✓ Names of the variables on each axis, and the units. The variable on the x-axis does not always have units.

✓ Choose a scale so that the chart is not squashed. The scale on the y-axis should end just after the highest value in your data. The scale should be easy to use, e.g. multiples of 1, 2, 5 or 10.

Worked example

Lauren drew this bar chart to show the different blood groups in her class.

Explain **two** improvements that Lauren could make to this bar chart.   **(2 marks)**

Label each axis so it is clear what information is being displayed.

Change the scale. There is a lot of wasted space in this chart as the last value on the y-axis is 14 but the units on this scale go up to 30.

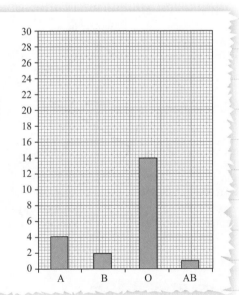

Now try this

1 A bar graph is shown on the right.

Describe **three** ways in which this bar graph could be improved.   **(3 marks)**

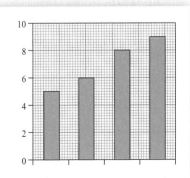

Pie charts

You need to be able to draw and interpret pie charts. These can be useful for data with 6 or fewer categories.

Worked example

MATHS SKILL

In a survey of class eye colour, 5 people have blue eyes, 15 have brown eyes, 8 have green eyes and 2 have hazel eyes.

Draw a pie chart to represent these data. **(3 marks)**

Total number of people = 5 + 15 + 8 + 2
                       = 30
Angle for 1 person = 360° ÷ 30 = 12°
Blue eyes: 5 × 12° = 60°
Brown eyes: 15 × 12° = 180°
Green eyes: 8 × 12° = 96°
Hazel eyes: 2 × 12° = 24°

Check: 60° + 180° + 96° + 24°
          = 360° ✓

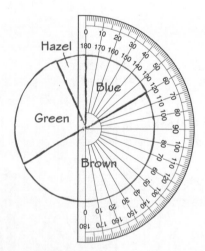

You need a sharp pencil, compasses and a protractor to draw a pie chart.

1. Work out the total number of people surveyed.

2. There are 360° in a full circle. There are 30 people. So divide 360° by 30 to find the angle that represents 1 person.

3. Multiply the angle that represents 1 person by the number of each eye colour to find the angle for each type.

4. Check that your angles add up to 360°.

5. Draw a circle using compasses. Draw a vertical line from the centre to the edge of the circle. Use a protractor to measure and draw the first angle (60°) from this line. Draw each angle carefully in order.

6. Label each sector of your pie chart with the correct eye colour.

Now try this

1 This pie chart shows the preferred snack choices from a college football team.

Describe why a bar chart would be a better method of presenting these data. **(2 marks)**

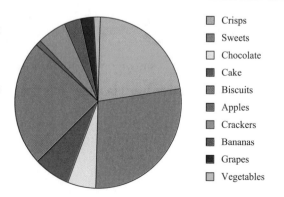

- ☐ Crisps
- ☐ Sweets
- ☐ Chocolate
- ☐ Cake
- ☐ Biscuits
- ☐ Apples
- ☐ Crackers
- ☐ Bananas
- ☐ Grapes
- ☐ Vegetables

Line graphs 1

LINE GRAPHS are useful to illustrate trends in data, usually how one variable varies over time.

Line graphs are usually used when:

Both the dependent and independent variables are quantitative, e.g. current or distance.

> The DEPENDENT VARIABLE is the outcome that is measured during the investigation.
>
> The INDEPENDENT VARIABLE is the variable that is controlled by the scientist.

Drawing line graphs

✓ Dependent variable always plotted on the vertical axis (y-axis).

✓ A title that provides a summary of the graph content.

Graph to show current through a resistor as voltage changes

(graph: Current (A) on vertical axis from 0 to 0.9; Voltage (V) on horizontal axis from 0 to 7; points plotted with crosses)

✓ Units with the axes' labels.

✓ Plot each point on the graph in pencil with a cross, so that the points are visible when a line is drawn through them.

✓ Choose scales so that your graph fills as much of the graph paper as possible. The scale should be easy to use, e.g. multiples of 1, 2, 5 or 10.

✓ Intervals between scale marks are at regular intervals.

✓ Independent variable plotted on the horizontal axis (x-axis).

Now try this

1 Carrie took part in a 30 minute exercise session. Once the exercise session was completed, Carrie measured her heart rate to see how long it would take to return to pre-exercise levels.

Time (minutes)	0	2	4	6	8	10	12
Heart rate (bpm)	125	110	102	95	80	72	70

Draw a line graph for these data.   **(6 marks)**

Make sure you include a title, label the axes and include units.

Line graphs 2

Using the correct axis

The axes of a line graph do not have to start at zero. If there are no results near zero the graph would have a lot of empty space and will not show the pattern clearly.

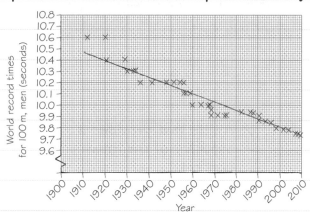

Different data

Several data series can be plotted on the same axes. This is useful to compare trends in different sets of data.

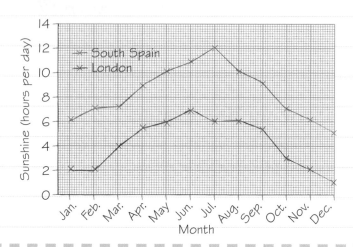

Worked example

MATHS SKILL

An investigation into exercise and recovery time gave the following results.

How long exercise lasted (min)	1	5	6	8
Time taken for recovery (min)	2	5	6	10

Plot a graph of the data. **(6 marks)**

There are improvements which could be made to this graph. The graph should have a title. The axes should be labelled and there should be units on the axes' labels. The intervals on the scale on the x-axis should also go up at regular steps.

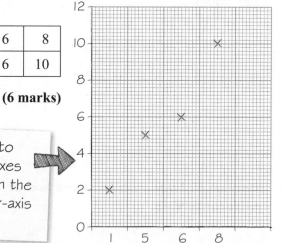

Now try this

1 A study into different diets collected the following results:

Month of study	1	2	3	4	5	6
Weight loss (kg/month)						
Diet A	3	1.5	1	1	0.5	0
Diet B	4.5	2	1	0.8	0	0.5

Make sure you include a title, label the axes and include units. You will need to plot this on a piece of graph paper.

Draw a line graph of these data. **(6 marks)**

Straight lines of best fit

To see a trend in data, you can draw a LINE OF BEST FIT through the points. This can show how the variables relate to each other.

Drawing a straight line of best fit

✓ Exclude any anomalous data.

✓ Draw the line so that about half the points are on either side of it.

✓ Use a ruler.

As the independent variable increases (the time), the dependent variable also increases (the temperature of the water). This can help to decide if the hypothesis is correct.

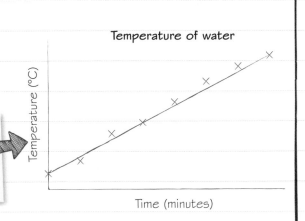

Temperature of water

Worked example

A line of best fit has been plotted to show the relationship between the two variables.

MATHS SKILL The data below is from an investigation into the effect of increasing drop height of a paper helicopter and the time taken to fall to the ground. Plot a graph of these data. **(6 marks)**

Drop height (m)	Time to fall (s)
2	1.2
4	2.3
6	3.6
8	4.9
10	6.1

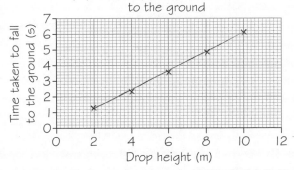

A graph to show how drop height of a paper helicopter affects time to fall to the ground

Now try this

1 The graph shows the results of an experiment to find out how the current through a resistor changed as the voltage across it increased. Complete the graph by drawing a line of best fit. **(2 marks)**

Remember to exclude any anomalous data from your line of best fit.

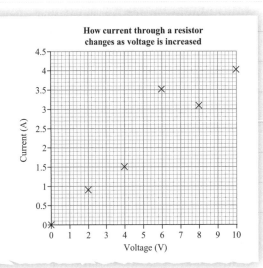

How current through a resistor changes as voltage is increased

Curves of best fit

When only two or three of the points seem to lie close to a straight line, a curve of best fit may show the pattern of the results better.

Drawing a curve of best fit

The line should be a smooth curve rather than a 'join the dots' approach! As with a line of best fit, the curve of best fit does not have to go through all of the data points.

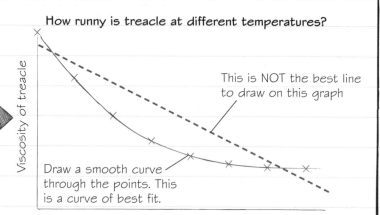

How runny is treacle at different temperatures?

This is NOT the best line to draw on this graph

Draw a smooth curve through the points. This is a curve of best fit.

Viscosity of treacle

Temperature (°C)

Worked example

MATHS SKILL

The graph below shows data points from an investigation. Draw a curve of best fit for these data points. **(2 marks)**

The question will state if you are supposed to plot a line of best fit or a curve of best fit.

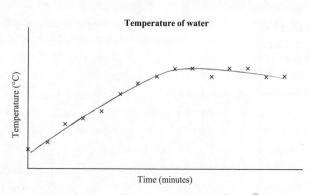

Temperature of water

Temperature (°C)

Time (minutes)

Now try this

1   The graph opposite shows the results of an investigation into current and voltage.

    (a) Complete the graph to show the correct line of best fit.   **(2 marks)**

    (b) Explain why you have chosen to use this line.   **(2 marks)**

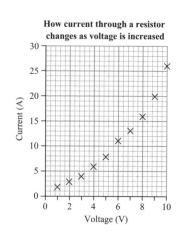

How current through a resistor changes as voltage is increased

Current (A)

Voltage (V)

Selecting types of graph 1

You need to be able to select the best type of graph to present any given set of data.

Bar chart

- Used to display qualitative or quantitative data.
- Used to compare differences between variables.

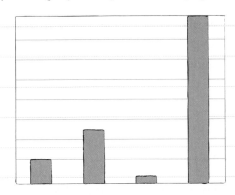

Line graph – straight line of best fit

- A line graph is used to display quantitative data and to show a trend in the variables.
- A line of best fit can be drawn to show a trend in variables.

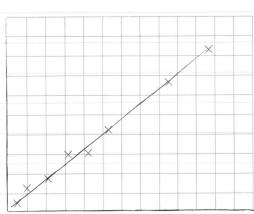

Line graph – curve of best fit

- A line of best fit can be a curve as well as a straight line. Both show a continual change in variables.

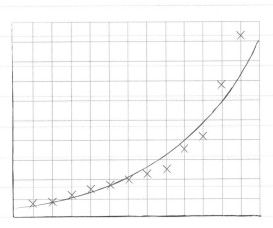

Now try this

1   For some fractions of crude oil, supply is greater than demand.

Fraction	Demand (%)	Supply (%)
petrol	40	20
paraffin	10	15
diesel	30	25
fuel oil	20	40

(a) Plot this information on a bar chart.   **(6 marks)**

(b) State why these data should be displayed on a bar chart rather than a line graph.   **(1 mark)**

Selecting types of graph 2

Pie chart

4

- Used to display qualitative or quantitative data.
- Maximum of 6 sets of data is best – more can be hard to see what the chart is trying to show.

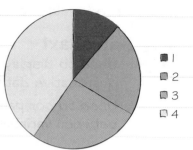

■ 1
■ 2
■ 3
■ 4

Scatter plot

5

- Used to display quantitative data.
- The data are plotted on a graph to see if there is a relationship between the variables.

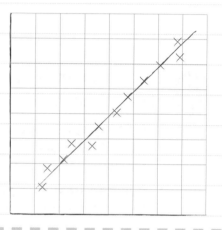

Worked example

George collected the data shown in the table opposite.

Explain which type of graph would be the best method to present these data.   **(3 marks)**

Both sets of variables are quantitative data, so a graph that uses this type of data would be best. As the dependent variable is increasing in line with the independent variable, a scatter plot with a line of best fit would be the best way of presenting this trend in the data.

Volume of water (ml)	Time to reach boiling point (min)
10	2
20	4
30	5
40	7
50	9
60	11
70	13
80	15

Now try this

1 Julie presents her data in this pie chart.

Describe which type of graph would be more appropriate to present these data.   **(2 marks)**

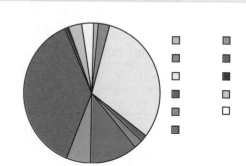

Finding values from graphs 1

It is possible to use a best-fit line on a graph to estimate other values that you did not test for within your data.

Worked example

Omar is investigating how his heart rate increases with exercise intensity. He exercises on a cycle ergometer and every 5 minutes the work load is increased at equal intervals of 10 kg.

This graph shows his results.

Estimate Omar's heart rate at a load of 25 kg. **(1 mark)**

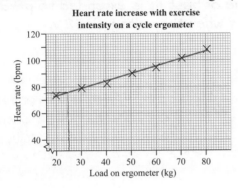

Heart rate increase with exercise intensity on a cycle ergometer

76 bpm

Follow these steps to estimate Omar's heart rate:

1. Draw a line of best fit on the graph using a ruler and a sharp pencil.

2. Use a ruler to draw a line up from 25 kg on the horizontal axis to the line of best fit.

3. Draw a line across to the vertical axis. Read your graph correct to the nearest small square.

Be really careful with the scale. On the vertical axis 1 large square represents 20 bpm, so one small square represents 2 bpm.

Always draw lines to show which values you are reading off your graph.

Worked example

A trampoline design engineer carries out an investigation to find out how a spring stretches in relation to the mass that is hung on it. A graph of her results is shown opposite.

Predict the extension of the spring at:

(a) 12 g   **(b)** 28 g   **(2 marks)**

(a) 0.6 cm   **(b)** 2.4 cm

Graph to show how a spring stretches with mass

Now try this

1 **(a)** Use the graph opposite to estimate the current at a voltage of 1.5 V.   **(2 marks)**

> Don't forget to give the units of the estimated answer.

(b) Explain how you could use the graph to estimate the current at a voltage of 8 V.   **(2 marks)**

Finding values from graphs 2

When the line of best fit is EXTENDED it is possible to predict results beyond the data in your investigation. This is called EXTRAPOLATION.

This can be useful in a conclusion to state what you think could have happened to the dependent variable if you had carried out further tests.

Be careful about making claims for data outside the range you have tested. Always say that the estimate may not be accurate because the trend may not continue.

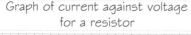
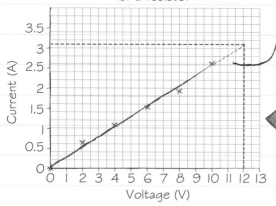

Graph of current against voltage for a resistor

The line of best fit is extended to predict the current when the voltage is 12 V.

Don't forget to ignore any anomalous data when drawing your line of best fit. Changing the slope of the line will have a significant effect on the accuracy of estimated values.

Worked example

MATHS SKILL

Luca investigates the effect of increasing the slope angle on the distance travelled by a marble rolled down the slope.

The results of the investigation are shown in the table opposite.

(a) Plot these results on a graph.   **(3 marks)**
(b) Using your graph, estimate the distance a marble would travel if the slope angle was at 5 degrees.   **(1 mark)**

Slope angle (degrees)	Distance (cm)
10	25.2
15	32.6
20	45.9
25	56.2

(a)

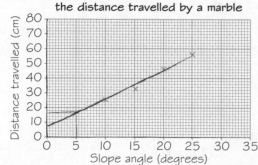

The effect of increasing slope angle on the distance travelled by a marble

(b) At 5 degrees the estimated distance that the marble will travel is 17 cm.

You can extrapolate a line backwards as well as extrapolating a line forwards.

Now try this

Use the information in the graph and table above to estimate the distance a marble would travel if the slope was at

(a) 30 degrees   **(1 mark)**   (b) 35 degrees   **(1 mark)**

Calculating gradients from graphs

You can calculate the GRADIENT of a graph to find additional information about the data.

Finding the gradient of a line

The gradient of a graph of distance (m) against time (s) is speed (m/s).

1. Draw a triangle from the line of best fit. Use a large triangle so your calculations are more accurate.

2. Find the change in y-axis values and change in x-axis values.

3. Gradient (slope) = $\dfrac{\text{change in } y\text{-axis value}}{\text{change in } x\text{-axis value}}$

4. The units will be the vertical axis units divided by the horizontal axis units.

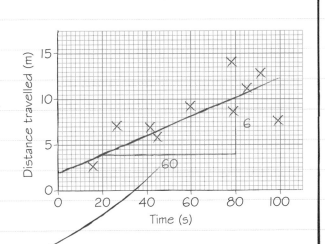

Gradient or speed = $\dfrac{6\,\text{m}}{60\,\text{s}} = 0.1\,\text{m/s}$

Worked example

MATHS SKILL

Use the graph of velocity (m/s) against time (s) to calculate acceleration (m/s²).   **(2 marks)**

Acceleration = $\dfrac{8}{4} = 2\,\text{m/s}^2$

In a graph of acceleration:
- horizontal line = an object is not accelerating, so moving with a constant velocity.
- line sloping upwards = an object is accelerating.
- line sloping downwards = an object is decelerating.

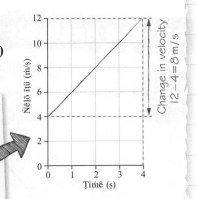

You could also be asked to calculate the gradient of a graph of voltage (V) against current (A). For a resistor at fixed temperature, this graph is a straight line and the gradient is the resistance in ohms (Ω).

Now try this

1   The graph opposite shows the velocity of a car from a stationary position. Use the gradient of the line to calculate the acceleration of the car between 4 and 10 seconds.   **(2 marks)**

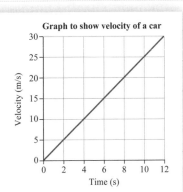

Calculating areas from graphs

You can calculate the AREA under a graph to find additional information about the data.

Velocity–time graphs show what the velocity of an object is over time. The gradient of the graph is equal to the acceleration of the object.

The area under the graph can be used to calculate the distance that the object has travelled.

1. Find the change in y-axis values and change in x-axis values.

   In this case the y-axis shows a change from 0 to 60 and the x-axis shows a change from 0 to 10. This means that the vehicle has changed velocity by 60 m/s over 10 s.

2. Multiply the two numbers together.

   Distance = 60 m/s × 10 s

3. The answer you are calculating is a distance so you need to use the distance unit that is part of the y-axis unit.

   Distance = 600 m

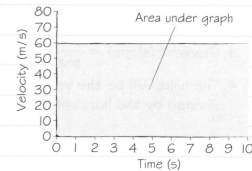

Worked example

MATHS SKILL

Shane plots a graph of the velocity of a train leaving a station.

Calculate the distance travelled by the train between 0 and 60 seconds.   **(3 marks)**

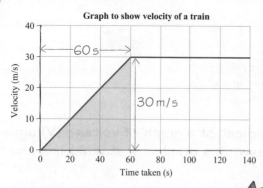

The area under the graph between 0 and 60 seconds is a triangle:

area of a triangle = $\frac{1}{2}$ × base × height

$(\frac{1}{2} \times 60) \times 30 = 900\,m$

Now try this

1 A car accelerates from 0 m/s to 30 m/s in 15 seconds. It travels for 10 seconds at 30 m/s and then decelerates to 0 m/s in 10 seconds.

   (a) Draw a velocity–time graph for the car's journey.   **(4 marks)**

   (b) Work out the distance travelled by the car as it decelerates   **(3 marks)**

Why anomalous results occur

An ANOMALOUS RESULT is a result that does not fit the pattern of the rest of your results or is out of the range of the other results.

Errors in taking measurements (human errors!) are the main reason why anomalous results occur in investigations, but occasionally the scientific equipment could have a fault.

Possible errors in an investigation

- Measurements taken incorrectly or not following the method and quantities correctly.
- Equipment not calibrated correctly.
- Not controlling some of the variables.
- Incorrect plotting of data on a graph.
- Incorrect recording of results.

It is always a good idea to repeat a part of an experiment if you spot an anomalous result. This will help to rule out human error.

It's possible your anomalous result might be important but you have not collected enough data to test this. So if you can't suggest a reason for why it doesn't fit the pattern:
- keep it in the table of results or graph
- exclude it from calculations and best-fit lines.

Worked example

A design engineer is investigating the effects of friction on movement. He rolls a trolley down a ramp and uses a stop watch to record the time taken to reach the end. He repeats the experiment, keeping the angle of the ramp the same but varies the friction by placing sandpaper of different grades of coarseness on the ramp. The results of the investigation are shown opposite.

Sandpaper grade	Time (s)
60	12.2
80	13.3
100	11.9
120	14.1

Describe why the result at a sandpaper grade of 100 could be an anomalous result.   **(2 marks)**

The higher the sandpaper grade, the higher the coarseness of the sandpaper, so the greater the friction it will create. The time taken for the trolley to roll down the ramp increases as the sandpaper grade increases due to the increased friction on the wheels of the trolley. The value at the sandpaper grade of 100 is the lowest of all of the values and does not increase in line with the rest of the values. As this value does not follow the pattern of the rest of the data, it could be an anomalous result.

Initial trolley position

Sandpaper

Sloping runway

Now try this

1 Max uses a digital balance to measure the mass of different substances. He forgets to set the scales to 0 before weighing each substance.
   Explain how this could produce anomalous data.   **(2 marks)**

43

Positive correlation

If two variables are CORRELATED there will be a clear pattern on the graph. The correlation may be due to chance, or there may be a true relationship. If there is a scientific explanation this means the change in one variable causes the change in the other.

Positive correlation

☑ As the independent variable increases, the dependent variable increases.

☑ Line slopes upwards.

For example, for a fixed resistance the current increases as the voltage increases.

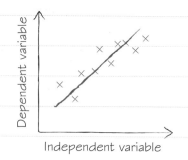

Worked example

Cathy investigated the relationship between height and body mass for a class of 30 students. She plots a graph of the results.

Describe the relationship between height and body mass for the class.   **(2 marks)**

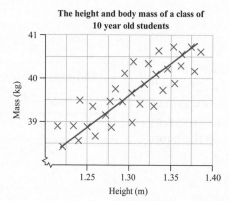

There are many points that are some distance from the line of best fit, so it is not possible to estimate accurately a student's mass from their height.

The taller students have a higher body mass. This is a positive correlation.

Remember: the closer the points are to a straight line, the stronger the correlation between the variables.

Now try this

1 A zoologist is investigating the relationship between the length and mass of a species of badger.

Describe the relationship shown on this graph.   **(2 marks)**

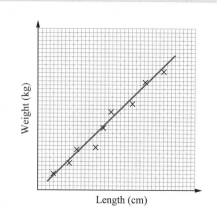

Negative correlation

Negative correlation

✓ As the independent variable increases, the dependent variable decreases.

✓ Line slopes downwards.

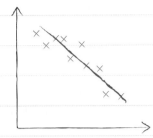

For example, an increase in physical activity is linked to a decrease in a person's body fat percentage.

Worked example

Charlie investigates the effect of shoe size on a person's IQ. The results are shown on this graph.

Describe the relationship shown by the results. **(2 marks)**

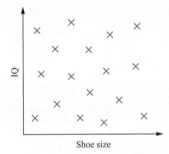

The data do not follow any pattern and a straight line or curve cannot be drawn on these data. Therefore, there is no correlation between these two variables.

Worked example

Cholesterol is linked to heart disease. A study into the level of cholesterol in the blood in men found the following:

Age	Mean blood cholesterol concentration (mmol/l)
15–24	4.5
25–34	5.0
35–44	5.4
45–54	5.8

Describe the relationship shown by the table. **(2 marks)**

As age increases the mean quantity of cholesterol in the blood increases. This is a positive correlation.

It is easier to spot a correlation from a graph, but if you look closely you can spot correlations from tables of data too.

Now try this

1 Suha is a nurse who specialises in lung diseases. She carries out an investigation into the effect of cigarettes on the volume of oxygen taken into the lungs.

Describe the relationship she finds shown on the graph.

(2 marks)

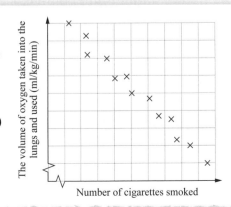

Direct proportion

A TREND is a pattern seen in tables of data or graphs. The shape of a graph will help you describe the trend.

When two variables are DIRECTLY PROPORTIONAL it means that when one changes, the other changes in the same way, by the same proportion or factor.

A directly proportional graph has:

☑ a straight line

☑ the line passing through the origin.

If a line of best fit does not go through the origin the variables are not directly proportional.

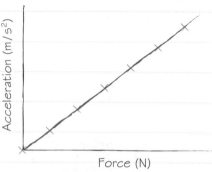

The effect of force on acceleration

This graph shows that acceleration is directly proportional to force.

> Direct proportion is a special case of positive correlation.

Worked example

Nazir is a fitness trainer. He carried out an investigation to find out the relationship between time and walking distance covered.

The results of the investigation are shown on the graph to the right.

Using the graph, describe the relationship between the two sets of data. **(2 marks)**

The graph shows that as time increases, the distance travelled also increases so there is a positive correlation.

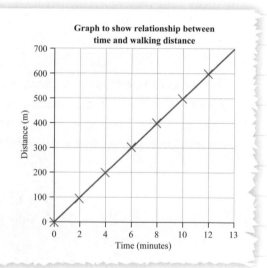

The answer isn't enough because it could also describe a curve. Since this graph goes through the origin on both axes, time is *directly proportional* to the distance travelled.

Now try this

1 (a) What is meant when two variables are said to be directly proportional? **(2 marks)**

 (b) Sketch a graph to illustrate variables that are directly proportional. **(4 marks)**

Inverse proportion

Inversely proportional

When one variable is INVERSELY PROPORTIONAL to another, one variable decreases as the other variable increases.

An inversely proportional graph shows:

☑ a curved line

☑ the line never passing through the origin.

This graph shows that current decreases as resistance increases, so current is inversely proportional to resistance.

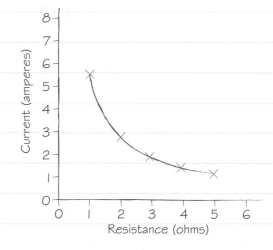

Worked example

Muneera, a physicist, carried out an investigation into the resistance of a thermistor at different temperatures.

Using the graph, describe the relationship between the temperature and the resistance. **(2 marks)**

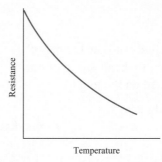

The graph shows that as the temperature increases the resistance through the thermistor decreases. This relationship is inversely proportional.

Now try this

1 The graph below shows the results from an investigation into the resistance of a light-dependent resistor.

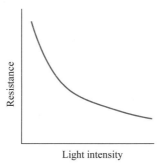

Describe the relationship shown in the graph between light intensity and resistance. **(3 marks)**

47

Using evidence to draw conclusions 1

You draw a CONCLUSION after you have analysed the evidence that you have collected. Here are some examples of good sentences for a conclusion:

From the mean temperature rise for each reaction and using the mass of each food item burned, I calculated the energy output per gram of food and this was highest for pumpkin seeds.

 Talk about any calculations you made.

The graph of voltage against current through the filament lamp shows that voltage increased as the current increased. However the graph curves upwards and is not a straight line, so the voltage is not directly proportional to the current and the resistance of the filament lamp is not constant.

 Write down any trends or patterns you have spotted in your data.

A conclusion should be based on all the known evidence.

Worked example

Martin carries out an experiment to find out how the pH of sodium hydroxide changes in relation to the quantity of nitric acid added to it. The results are plotted on the graph.

Write a conclusion that Martin could make for this investigation.   **(4 marks)**

There is a negative correlation between the quantity of nitric acid added to sodium hydroxide and the pH of the solution.

As more nitric acid is added to sodium hydroxide, the pH of the solution becomes lower. This is because nitric acid is an acid and has a low pH. Sodium hydroxide is a base so has a higher pH than the acid. As more acid is added to the sodium hydroxide it starts to neutralise the sodium hydroxide which, at about 35 ml, brings the pH to neutral. As more nitric acid is added the solution starts to turn acidic, which results in the lower pH.

pH of sodium hydroxide when nitric acid is added

(graph: pH on y-axis from 0 to 14, Nitric acid (ml) on x-axis from 0 to 60, showing a downward trend line)

Now try this

1 Jess carries out an investigation into how the temperature of a cold pack used to treat sports injuries changes over time. She measures the temperature of the cold pack over a 10 minute period.

Describe the conclusion that can be drawn form this graph.

(2 marks)

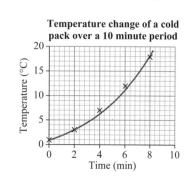

Temperature change of a cold pack over a 10 minute period

(graph: Temperature (°C) on y-axis from 0 to 20, Time (min) on x-axis from 0 to 10)

Using evidence to draw conclusions 2

Worked example

Dwaine is a chemist. He carries out an investigation to find out how the temperature of dilute hydrochloric acid changes in relation to the volume of sodium hydroxide added to it.

The results of the investigation are shown on the graph.

In his conclusion, Dwaine states that the higher the concentration of sodium hydroxide, the greater the change in temperature of the hydrochloric acid.
Can this conclusion be drawn from these data? **(2 marks)**

No. This is because Dwaine is investigating the effect of changing the volume of sodium hydroxide. He is keeping the concentration fixed.

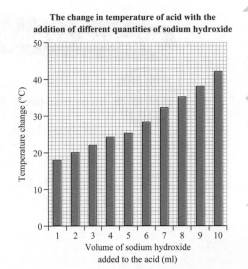

The change in temperature of acid with the addition of different quantities of sodium hydroxide

Temperature change (°C) vs Volume of sodium hydroxide added to the acid (ml)

Worked example

Explain if the chart that Dwaine used to present the results is the best type of chart to present these data. **(2 marks)**

A bar chart has been used to present these data but a line chart would have been better, as the results are showing that there is a relationship between the variables. Bar charts are often used to display qualitative and quantitative data rather than just quantitative data.

Bar charts are usually used when the independent variable is qualitative data and the dependent variable is quantitative data.

Line charts are usually used to illustrate data when both the dependent and independent variables are quantitative data. They help to show a relationship between the variables.

Now try this

1 Use the evidence presented in the graph above for Dwaine's investigation to draw a suitable conclusion. **(3 marks)**

Learning aim B sample questions

This page looks at the type of questions you might be asked to answer as part of the Unit 8 Scientific Skills paper. These questions cover material from Learning aim B. The experiment investigates how temperature affects the rebound height of a squash ball and carries on from the example used on page 22.

Worked example

MATHS SKILL

Results

Temperature (°C)	10	15	20	25	30	35	40
Rebound height (cm)	80	82	84	85	83	90	91

Plot a graph of the data. **(6 marks)**

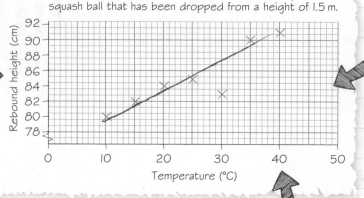

The effect of temperature on the rebound height of a squash ball that has been dropped from a height of 1.5 m.

Use the right labels on the axes, including the units. Make sure you have used the right scale too – your graph should fill the graph paper.

Join your points and look closely at the pattern of the points. If they are in a straight line, then use a line of best fit. If they fall on a curve, then use a smooth curve of best fit. If there are any points that look like they fall off the line, then they may be anomalous so leave them off the line.

Plot the points carefully.

Worked example

Explain why the experiment should be carried out more than once. **(2 marks)**

This makes it easier to see the anomalous results. This means that any anomalies can be ignored in calculations and when plotting a graph. It could also mean the anomalous results could be repeated. This helps to make the results more reliable.

This is an excellent answer because it explains why carrying out an experiment more than once makes the results more reliable.

Now try this

1   An investigation was carried out to find out how fast 6 people could run 100 m. The results are shown on the right.

Plot a bar chart for these data. **(6 marks)**

Runner	Time (s)
1	15.5
2	12.6
3	11.2
4	13.4
5	14.8
6	30.2

Writing a conclusion 1

The conclusion is a summary of an investigation and discusses what has been discovered and why it has happened.

 Restate the hypothesis

What did you predict would happen?

> It was predicted that the larger the surface area of the parachute the longer it would take to fall to the ground.

2 Accept or reject your hypothesis

State if your results support your hypothesis.

👍 Results support hypothesis – accept it.

👎 Results do not support the hypothesis – reject it.

> The results from this investigation support this hypothesis.

 Use data to confirm the hypothesis

Give results or data that confirm why you have accepted or rejected your hypothesis.

> The results showed that the parachute with the largest surface area took the longest time to fall to the ground and the parachute with the smallest surface area took the shortest time to fall to the ground.

 State the relationship

Was there any type of relationship between the variables in your investigation? How did the independent variable affect the dependent variable?

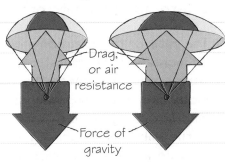

Drag, or air resistance

Force of gravity

> There was a positive correlation between the surface area of the parachute and the time it took to fall to the ground.

 Use scientific theory to justify your results

> The force of gravity pulls objects down towards the ground. The parachute causes air resistance, which pushes the parachute back up and creates an opposite force to the force of gravity. The larger the surface area of the parachute, the more air resistance is created which slows the fall more than parachutes with smaller surface areas.

 Use a concluding sentence to sum up the results of your investigation

> Therefore, in conclusion, it can be stated that parachutes with larger surface areas take longer to fall to the ground compared with parachutes with smaller surface areas.

Now try this

1　Identify **six** factors that should be included in a conclusion.　　　(6 marks)

Writing a conclusion 2

Look at this example of a good conclusion and then read the post-its.

This experiment was investigating the effect of different temperatures on the activity of the enzyme sucrase in breaking down sucrose to glucose.

The hypothesis was that as the temperature increases, the rate of enzyme reaction will increase.

1. The hypothesis is restated here.

The results prove my hypothesis, so the hypothesis is accepted.

2. The hypothesis is accepted.

The results show that there was a positive correlation between the temperature and the rate of reaction. As the temperature increased from 5°C to 30°C, the rate of reaction increased.

3. Data is provided to confirm why the hypothesis has been accepted.

4. The relationship between the variables is stated.

The rate of the reaction increased in-line with an increase in temperature, because the rise in temperature increases the kinetic energy of the molecules in the substances. This increase in kinetic energy results in more of the molecules in the substances coming into contact with the enzyme and colliding with its surface with greater energy and so increasing the rate of reaction.

5. Scientific theory is provided to justify the results.

Therefore, to conclude, my investigation showed that there was a positive correlation between temperature and the rate of the reaction of the enzyme sucrase.

6. The concluding sentence sums up the results of the investigation.

Now try this

1   An investigation was carried out into the relationship between the acceleration of a trolley travelling down a sloped ramp, and the force applied to the trolley. The results are shown below.

Force (N)	0.2	0.4	0.6
Acceleration (m/s²)	1.26	2.41	3.82

Write a conclusion for this investigation.   **(5 marks)**

Inferences

An INFERENCE is a conclusion based on limited information or evidence. If your first inference is incorrect you might need more information to move from an inference to a conclusion.

Drawing inferences from conclusions

Scientific investigations often follow a cycle of inferences and conclusions:

An inference usually leads to the development of another hypothesis. This creates further investigations that test if the new hypothesis can be proved.

Uses data to make a conclusion → Makes an inference from the conclusion → Decides whether hypothesis is correct → Develops hypothesis → Devises way of testing hypothesis → Scientist gathers data →

Worked example

Jacob is a marine biologist. He carries out an investigation into the behaviour of stickleback fish. He found that male stickleback fish that have red undersides attack each other.

He then placed models of fish with red undersides into the pond as well as models without red undersides. He found that the models of fish with red undersides were attacked by the male sticklebacks with red undersides, but the models without red undersides were not attacked.

Describe an inference that can be drawn from this investigation. **(2 marks)**

Male stickleback fish with red undersides will only attack fish with red undersides. This may be due to the fact that the males turn red to attract females. They then attack other fish with red undersides as they think that they are competition to mate with the female fish.

Now try this

The table opposite shows the number of fish species found at different depths in the sea.

1 Make an inference related to sea depth and the number of fish species, based on the information provided in the table. **(2 marks)**

2 A student looked into a freshly dug hole in the ground and wrote down statements about the sediment at the bottom of the hole. Which of the following statements is an inference? **(1 mark)**

A ☐ The sediments were deposited by a stream.

B ☐ Over 50% of the sediments are the size of sand grains or smaller.

C ☐ Some of the particles are rounded.

D ☐ The hole is 2 metres deep.

Depth (m)	Fish species
25	35
50	32
75	31
100	22
125	15
150	13

The inference will be something that the student has not been able to see or find out for themselves from looking in the hole but something that they think has happened.

Support for a hypothesis 1

You need to be able to decide if there is enough good evidence to show that the hypothesis is correct.

1 **Restate the hypothesis**

What did you predict would happen?

Doubling the area of the parachute will double the time taken to fall to the ground.

3 **Use data to confirm the hypothesis**

The parachute with the largest surface area of 4900 cm² took the longest time to fall to the ground (4.2 s).

Also the parachute with the smallest surface area of 100 cm² took the shortest time (1.7 s) to fall to the ground.

2 **Accept or reject the hypothesis**

Look at the results you gathered.

👎 Results do not support the hypothesis – reject it.

Surface area (cm²)	100	400	900	1600	2500	4900
Mean time to fall 2 m (s)	1.7	1.9	2.3	3.1	3.6	4.2

The results from this investigation do not support the hypothesis.

Worked example

James carried out an investigation to determine the effect of wind on the rate of transpiration in plants. His hypothesis was that the wind increases the rate of transpiration compared with normal non-windy conditions. Water loss by the plants was recorded at 10-minute intervals for 30 minutes.

Using the results collected in the table, describe if this hypothesis can be supported. **(2 marks)**

	Mean water loss (ml)			
	0 min	**10 min**	**20 min**	**30 min**
No fan	0.0	2.1	4.5	6.5
With a fan	0.0	4.3	7.7	11.8

Yes, the hypothesis can be supported by these data. The water loss is much greater with a fan over time – after 30 minutes the water loss is 11.8 ml which is almost twice as much with a fan than without a fan at 6.5 ml.

Now try this

A BMX park designer stated this hypothesis: the greater the angle of take-off on a jump, the further the distance a bike will travel. The results of the investigation are shown in the table below.

Slope angle (degrees)	30	40	50	60	70
Distance travelled (cm)	457	541	678	620	572

Using the information collected, describe if this hypothesis can be supported. **(3 marks)**

Support for a hypothesis 2

Once you have decided whether you can accept or reject the hypothesis you need to work out if there is a relationship between the variables, using your scientific knowledge.

 State the relationship

Look at the graph of your data. How did the independent variable affect the dependent variable?

There was a positive correlation between the surface area of the parachute and the time it took to fall to the ground. However, the graph does not show a directly proportional relationship, so the quantitative part of my hypothesis is not supported.

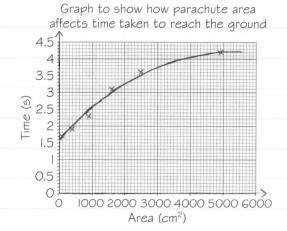

Graph to show how parachute area affects time taken to reach the ground

5 **Use scientific theory to support the conclusion**

6 **Use a concluding sentence to sum up the results of your investigation**

The object's weight pulls it towards the ground. The parachute causes air resistance, which creates a force in the opposite direction to the force of gravity.

The larger the surface area of the parachute, the more air resistance is created. This is due to more air particles colliding with the parachute, which slows the fall more than parachutes with smaller surface areas.

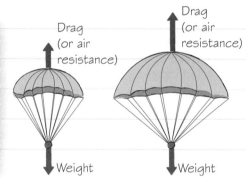

Worked example

The table below shows the relationship between age and level of physical activity for people aged between 5 and 30. Describe the relationship between age and level of physical activity.   **(2 marks)**

Age range	5–10	11–15	16–20	21–25	26–30
Average level of physical activity (hours per day)	1.75	1.25	0.75	0.5	0.5

There is a negative correlation between age and physical activity, as a person's age increases, the average level of physical activity decreases.

Now try this

1   The table shows the results of an investigation into the relationship between force, mass and acceleration for a trolley travelling down a ramp. The mass was kept constant (control variable).

(a) Plot a graph of these data.   **(6 marks)**

(b) Use the graph to state the relationship between the force and acceleration.   **(2 marks)**

(c) Use your scientific knowledge to explain these results.   **(3 marks)**

Force (N)	Acceleration (m/s²)
1.0	0.24
2.0	0.50
3.0	0.73

Support for a hypothesis 3

You need to be able to explain how the evidence that has been presented supports a hypothesis.

Worked example

The results from an experiment to investigate the effect of wind on the rate of transpiration in plants are shown below.

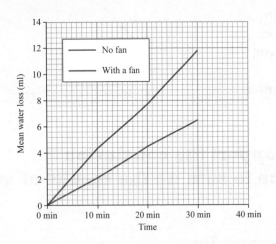

Plant cutting
Air-tight seal
Calibrated pipette
Water-filled tube
Potometer

Explain if this evidence supports the hypothesis: the effect of wind increases the rate of transpiration in a plant compared with normal classroom conditions. **(5 marks)**

The hypothesis was that the effect of wind increases the rate of transpiration in a plant compared with normal classroom conditions.

The results do support the hypothesis, so the hypothesis is accepted.

The results show that at 30 minutes, a mean of 11.8 ml of water was transpired from the plants by the effect of wind compared with a mean of 6.5 ml with no wind.

The reason the rate of transpiration was greater with the effect of wind compared with no wind was because wind increases evaporation of water from leaves. This has the effect of increasing the movement of water into the roots, which is then drawn up the leaves of the plant and so increases the rate of transpiration.

> Restate the hypothesis in your answer so that you can refer back to it.

> Then, based on the results that you have seen, state if the hypothesis should be accepted or rejected.

> Include data from the results to confirm why the hypothesis has been accepted.

> Scientific theory should then be provided to justify the results.

Now try this

1 In a specific variety of corn, the kernels turn red when they are exposed to sunlight. When they are not exposed to sunlight, their kernels remain yellow.

Which of the following hypotheses is supported from this evidence? **(1 mark)**

A ☐ A different type of DNA is produced by the corn kernel when sunlight is present.

B ☐ A different species of corn is produced in sunlight.

C ☐ The effect of sunlight affects the number of chromosomes inherited by a corn kernel.

D ☐ Sunlight has an effect on gene expression in some types of corn kernel.

Strength of evidence

A conclusion should discuss how well the evidence supports the conclusion.

Strong correlation

A line graph of your results will help you discuss the strength of your evidence.

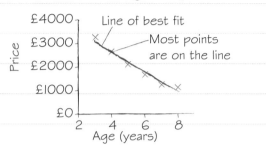

Strong negative correlation between the variables.

Weak correlation

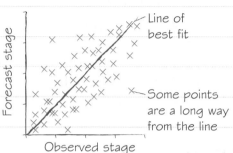

Very weak positive correlation between the variables.

If the evidence does not allow you to be certain about the conclusion, the hypothesis is not supported by the evidence.

If the evidence shows very limited support or no support for the hypothesis, it may be that you can only draw inferences from the conclusion.

Another investigation might be needed to gather more data.

Worked example

Researchers at a hospital carried out an investigation to find out if eating more fruit and vegetables every day decreases the risk of bowel cancer.
The results of the investigation are shown below.

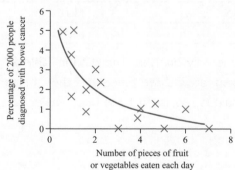

Explain how well the evidence supports this hypothesis. **(2 marks)**

There is a weak negative correlation between the percentage of people diagnosed with bowel cancer and the quantity of fruit and vegetables eaten each day – the more fruit and vegetables that are eaten, the less chance there is of a person being diagnosed with bowel cancer. However, a number of points are quite some distance from the best fit curve and very few points are on the best fit curve, so the evidence only provides weak support for this hypothesis.

You need to be able to comment on **how well** the evidence supports the conclusion, based on **all** the evidence provided. Say what relationship the graph shows, and comment on whether the correlation is weak or strong.

Now try this

1 Describe what you would expect to see on a line graph of results if the results:
   (a) strongly supported the hypothesis
   (b) weakly supported the hypothesis
   (c) showed no support for the hypothesis.   **(3 marks)**

Sufficient data 1

When evaluating an investigation you need to be able to say if enough data points were taken and if the data taken were reliable enough to draw a valid conclusion.

Repeating measurements

An investigation should include repeating measurements at least three times to check that the results are REPEATABLE.

- If the values vary slightly a MEAN can be taken – the mean of several measurements is more accurate than a single value.
- If one of the measurements is very different (ANOMALOUS), this is easy to identify.

Worked example

Comment on the reliability of the data shown in Table 1 and Table 2. **(3 marks)**

Table 1

Concentration of HCl (M)	Temperature change (°C)
1	5
2	6
3	8
4	12
5	18

It is important to identify anomalies because they can then be excluded from calculations. Anomalies can make you come to the wrong conclusion.

Table 2

Concentration of HCl (M)	Temperature change (°C)	Mean temperature change (°C)
1	5, 4, 4	4.3
2	7, 6, 6	6.3
3	8, 8, 7	7.7
4	10, 12, 11	11.0
5	14, 13, 18	13.5

The data in Table 1 is not very reliable because there is only one set. The data in Table 2 is more reliable because each experiment has been repeated three times and then a mean calculated. It is difficult to see if there are any anomalous results in Table 1 but easier to spot anomalies in Table 2.

Now try this

1 Gareth is carrying out an investigation to find the power of a person's legs. The person stands next to a wall, reaching up as high as possible with their feet flat on the floor and makes a chalk mark on the wall at the highest point that they can reach. They then jump as high as possible and make a second chalk mark on the wall at the highest point of the jump.

In the investigation, each person performs the test once.

Explain why carrying out the test once will not provide sufficient data for a valid conclusion to be drawn. **(2 marks)**

Sufficient data 2

When evaluating an investigation you need to be able to say if the measurements cover a wide enough RANGE.

Range of measurements

A bigger range of results will help find a pattern or trend.

When selecting the values of the independent variable, include at least six different values, evenly spaced over the range.

Worked example

Janis is investigating the resistance of a bulb in a set of Christmas tree lights. She measures the current when three different voltages are applied.

Janis comes to the conclusion that there is a strong positive correlation between the voltage and current through the Christmas tree lights.

Explain if these data provide sufficient information to draw a strong conclusion.

(2 marks)

Three points on a graph is not enough to be sure of the relationship – there is not enough information to draw a strong conclusion from this set of data. There need to be six or more data points to provide sufficient evidence for a strong conclusion.

An investigation should include repeating measurements at least three times to provide reliable data.

Reliable data

If the data collected in an investigation are reliable, it means that if another person carried out your investigation they would obtain the same, or very similar results. This makes the investigation repeatable.

Data can be unreliable if there are issues with the quality of the data such as:

- imprecise data
- subjective observations.

Subjective observations are an individual's feelings, thoughts or opinions and will differ between different people.

For example, how hot something feels or how sweet something tastes will vary depending upon the person who is performing the test, so the data are not very reliable.

Now try this

1 A sweet manufacturer carried out an investigation into how eating chocolate made people feel. The people surveyed had to say how they were feeling before eating the chocolate, two minutes after eating the chocolate, and again five minutes after eating the chocolate.

The people taking part in the study had to rate how they felt from the following options: very sad, sad, ok, happy, very happy.

Explain if the data collected in this experiment would be reliable.   **(3 marks)**

Validity of a conclusion 1

A conclusion is only VALID if it is based on accurate results and a correct scientific method.

Was the experiment relevant to the question being asked?

Is this conclusion valid?

Was the experiment set up in a sensible way, and was the correct method used?

Is the conclusion based on the results, or has it been influenced by other ideas?

Worked example

A microbiologist carries out an investigation using antibacterial soap and non-antibacterial soap.

His hypothesis is that the soap with antibacterial ingredients will kill more bacteria compared to non-antibacterial hand soap. He places 5 ml of each soap into a Petri dish containing *E. coli* cultures and leaves them for 3 days.

He then measures the diameter of a bacterial colony around the sample of each soap. The results are shown below.

Type of soap	Bacterial growth (μm)
Antibacterial	20
Non-antibacterial	21

He draws the conclusion that antibacterial soap should be used to wash people's hands as this type of soap kills more bacteria compared to non-antibacterial soap.

Explain why this not a valid conclusion.   **(2 marks)**

This conclusion is not valid. The experiment tests how well different soaps prevent bacterial growth, but the conclusion is talking about the growth of bacteria on hands and how you wash your hands, and this was not what was tested in the experiment.

The results from this experiment are not significant either, as there is only 1 μm difference in the diameter of the bacteria colony. This difference could be the result of human error or variation in the method.

Now try this

1   A gardener carried out an experiment to test the effect of light on plant growth. His hypothesis was that sunlight would increase the rate of plant growth by 50% compared with artificial light.

The plants were grown in separate pots; one pot was exposed to sunlight, the other to artificial light. All other conditions were kept the same. The height of each plant was measured at the start and after a 3 month period.

Type of light	Sunlight	Artificial light
Increase in plant height (cm)	5.0	4.5

The gardener concluded that all plants grow more rapidly in sunlight than in artificial light.

Explain if this conclusion is valid.   **(3 marks)**

Points to consider:
* is there a significant difference in height for each plant?
* were there enough individual plants used in the experiment?

Evaluating an investigation 1

EVALUATING means assessing how good or how bad something is. You must give reasons for what you say, and then comment on what you could have done to improve it.

Evaluating the method

- How could you improve the method to give better evidence? Explain why each change would give better-quality data.
- Were there any weaknesses in the method? Was it difficult to get good-quality data?
- Look back at the method for an explanation of any anomalous results.

Remember, an anomalous result may also mean that a new scientific discovery has been made!

Worked example

A biologist investigated how herbs affect bacterial growth. He washed 4 different types of herb in water, then placed a measured quantity of each type of herb into a Petri dish containing *E. coli* cultures and left them for 3 days.

He then measured the diameter of any bacterial colony around the sample of each herb.

In his evaluation he writes, 'the results for coriander seem to be anomalous as the measurement for bacterial growth was much higher than for other herbs. I am not sure why this happened.'
Explain how this section of the evaluation could be improved.   **(3 marks)**

He should have tried to explain why there was the anomalous result and referred back to his method to get some ideas on what may have happened. It could be that the coriander was not washed properly and there were some microorganisms on the coriander from the soil it was grown in. This would then be responsible for this anomalous result.

An anomalous result usually means that an error has occurred in the experimental processes – readings or measurements may not have been taken accurately, which has resulted in this 'odd' piece of data.

Now try this

1 Simon is carrying out an investigation to find out how burning different masses of ethanol affects the temperature of water.

He adds different masses of ethanol to an ethanol burner. He records the temperature of the water, then lights the ethanol; when the ethanol has finished burning, he takes the temperature of the water again and records the temperature change.

His results are shown opposite.

Give **two** possible errors that Simon could have made in his experiment to produce the anomalous result at the 10 g mass of ethanol.   **(2 marks)**

Mass of ethanol (g)	Temperature increase (°C)
2	2
4	5
6	7
8	11
10	11
12	14
14	16

Evaluating an investigation 2

The conclusion should also be evaluated based on the data that has been collected.

Evaluating the conclusion

- Did the data you collect give you enough information to draw a conclusion?
- Is the conclusion valid?

This is an example of a good evaluation:

The results showed that I could accept my hypothesis that the thicker the piece of wire, the lower the resistance.

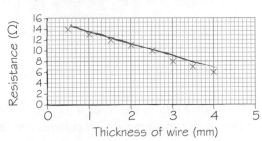

I was able to draw a valid conclusion as I collected enough data to find a relationship between the thickness of a piece of wire and its resistance. I know that there was enough data because the best-fit line passed close to all the points of data that I collected.

These results showed that there was a negative correlation between the variables – as the thickness of the wire increased, the resistance decreased.

Worked example

A student performed an experiment to determine if treating 500 tomato plants with a plant growth hormone will make them grow faster. The results are shown in the table below.

Days	Height (cm)
1	5.5
3	6.0
5	6.2
7	6.5
9	7.0
11	7.3

She concludes that the plant hormone makes tomato plants grow faster than if they did not receive the hormone.

Explain why this is not a valid conclusion from these results. **(2 marks)**

The method is flawed, as there is no control group to compare the group that have received the hormone with a group that have not received the hormone. Therefore, this is not a valid conclusion.

 Only one set of data has been obtained for each value, which means that the data may not be reliable.

Now try this

1 A sports psychologist is carrying out an investigation to find out about the effects of cooling on a person's reaction times. They carry out a ruler-drop experiment that tests reaction times with five males aged between 18 and 19 years old.

Test 1: The person places their hands in a water bath at a temperature of 40 °C for 2 minutes.
Then carry out the reaction test once and record their score.

Test 2: The person cools their hands in a water bath at 5 °C for 2 minutes.
Then carry out the reaction test once and record their score.

Give **one** reason why the data from this experiment may not be reliable. **(1 mark)**

Improving an investigation 1

There are always ways to improve an investigation. You should suggest and justify (give reasons for) how the method could be improved to give more accurate and reliable results.

Ways to improve an investigation

- ✓ increase the range of the independent variable
- ✓ choose different measuring equipment, e.g. a more accurate timer
- ✓ choose a different method to measure the dependent variable
- ✓ consider if all the variables were controlled
- ✓ consider how the variables can be better controlled.

Repeating measurements

It can be a good idea to repeat the measurement several times for the same set of variables and then calculate a mean. Anomalous results will still occur, but by working out a mean the data will be closer to the true value.

For example, measuring the rebound height of a ball is difficult as the ball moves quickly and is only at the top of the rebound for a split second. Measuring a number of rebound heights for each drop height and calculating the mean will give a more accurate value.

Worked example

Max is trying to work out what chemicals would be best to use in a hand warmer. He carries out an investigation using different masses of ammonium chloride and adding this to 30 cm³ of water.

He records the temperature of the water first. He then measures out different masses of ammonium chloride on digital scales and adds it to the water. He mixes the water with the ammonium chloride. By having a thermometer placed inside the beaker, he takes the reading when it has reached its highest temperature.

The results of the investigation are shown below:

Mass of ammonium chloride (g)	Temperature rise (°C)
3	4
9	7
15	11

Suggest a way to improve this investigation.   **(3 marks)**

Increasing the range of the independent variable (the mass of ammonium chloride) as this will help to show more clearly any trends and relationships between the independent and dependent variables. This investigation only includes three levels of the independent variable, so there is insufficient data to really show any trends or relationships between the variables.

Make sure the improvement you suggest is practical, and you should also consider how this improvement will add to the conclusion.

Now try this

Remember, repeating measurements helps to identify anomalous results!

1   Explain why repeating measurements can be a good way to improve an investigation.   **(2 marks)**

Improving an investigation 2

You need to be able to look at a method or part of a method and give some ideas on how to improve the range or quality of the data.

Increasing the sample size

The SAMPLE SIZE is the number of data points collected. To increase the sample size you can increase the number of values within the range (smaller intervals) or extend the range.

Repeating experiments

Repeating the whole experiment again using the same method and comparing the results helps the investigator (or another scientist) to see if the results they obtained are repeatable.

Difficult to find a trend in the data so conclusion may not be valid

More chance of getting meaningful results

SMALLER   SAMPLE SIZE   LARGER

Sampling error tends to be larger

More statistically significant results

Worked example

An environmental scientist is investigating the temperature of water at different depths in a pond that is 10 m wide, 30 m long and 1.5 m deep. His hypothesis is that the deeper the water, the lower the temperature. He takes 2 samples of pond water, one at depth of 0.5 m and one at 1.0 m. His results showed that the deeper the water, the lower the temperature.

Explain **one** change to this method that would improve his results.   **(3 marks)**

The pond is very large and he has only taken samples at two depths. He should increase the number of measurements at different depths. This will increase the range of data, so there is more chance of seeing a pattern in the results.

To give the best possible answer, make sure you describe the improvements in detail and also explain why they would be an improvement.

Now try this

1 Alison carried out an experiment to find out how the mass of salt added to water affected the boiling point. The results are shown in the table.

Mass of salt (g)	Boiling point (°C)
10	95
20	85
30	72

Suggest an improvement to the experiment.   **(3 marks)**

Extending an investigation

You need to be able to suggest ways to extend your investigation to provide stronger support for the hypothesis, or to test a related hypothesis.

Extending the original hypothesis

If you reject your hypothesis, there may still have been trends in the results that you need to explore further by changing the design of the experiment.

For example:
- using more values of the independent variable
- changing the conditions of the investigation
- changing the sample being tested.

Testing a related hypothesis

You could also suggest testing the effect of changing a different variable:
- use a new but similar independent variable
- remember to control other variables so that you are testing the effect of changing only one variable.

Worked example

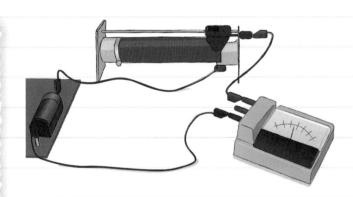

Amir used a voltmeter to test how the output voltage of alkaline and non-alkaline batteries changed over time for different currents through a resistor in the circuit.

He found that for all currents, the alkaline battery maintained its output voltage longer than the non-alkaline battery.

Amir suggested that temperature may also affect the batteries output voltage.

(a) State an appropriate hypothesis for this new experiment. **(2 marks)**

(a) If the surrounding temperature of the battery is increased this will decrease the output voltage of the battery. The non-alkaline batteries will decrease their voltage more quickly than alkaline batteries, because they performed worse in the first experiment.

(b) Suggest how Amir could extend the investigation to test this hypothesis. **(2 marks)**

(b) He must give a suitable method that explains how it will test the new hypothesis, not the old hypothesis e.g. the independent variable is now temperature, so the current in the circuit needs to be kept constant.

Now try this

1 Beth investigates a hypothesis that the greater the drop height of a ball, the higher the rebound height. She used three different types of ball – a tennis ball, a table tennis ball and a basketball. She found in each case, the greater the height the ball was dropped from, the higher the rebound height.

Describe an idea that could be used to extend this project. **(2 marks)**

Describe the evidence you might collect to extend the investigation and say why you would collect it.

Learning aim C sample questions

This page looks at the type of questions you might be asked to answer as part of the Unit 8 Scientific Skills paper. These questions cover material from Learning aim C. The experiment investigates how temperature affects the rebound height of a squash ball and carries on from the example used on pages 22 and 50.

Worked example

Explain whether the results supported the hypothesis or not. **(3 marks)**

The graph shows that there is a positive correlation between the increase in temperature of the squash ball and the height achieved on the rebound. This means I should accept the hypothesis, which stated that the bounce height would increase with temperature.

The conclusion of this experiment was that the temperature of the ball did affect the height of the rebound.

 The graph of the results can be found on page 50.

 This is a good answer because it clearly says whether the hypothesis can be accepted or rejected and then explains why.

Worked example

Explain why taking more readings for each temperature interval and calculating a mean would help to improve the reliability of the results. **(3 marks)**

Measuring the rebound height of a ball is difficult, as the ball moves quickly and is only at the top of the rebound height for a split second. By taking a number of rebound measurements at each drop height and calculating the mean, it will give much more reliable results.

This is a good answer because it explains why taking more readings increases the reliability of the results.

The answer could also explain that when working out the mean, any anomalous results should be left out of the calculations, as these can have a significant effect and make the results less accurate.

Now try this

1 Suggest how this experiment could be extended. **(2 marks)**

Testing the internal pressure of each squash ball at different temperatures may be a good idea, as an increase in the internal pressure of the ball may be the reason why there is an increase in rebound bounce height.

Answers

Learning aim A

1. Scientific equipment 1

1
A Stirring rod = 2 Stirs the liquid without becoming hot.
B Bunsen burner = 3 Source of heat.
C Conical flask = 1 Holds the liquid that is being heated.

2. Scientific equipment 2

1 The clamp stand will keep the test tube held at the same distance above the Bunsen burner flame to make the test fair. Using a clamp and clamp stand will also reduce the risks of holding the test tube over the Bunsen burner, e.g. being burnt by the Bunsen burner or the hot water, or dropping the test tube.

3. Chemistry equipment

1 (a) a pipette
(b) a measuring cylinder
(c) a thermometer
(d) a pH meter.

4. Physics equipment

1 She will need:
- a Newton meter to measure the force needed to push the trolley
- a tape measure to measure the distance covered

5. Biology equipment

1 (a) a microscope
(b) a glass slide
(c) a cover slip
(d) a Petri dish containing nutrient agar
2 She will need:
- some form of exercise equipment, e.g. a treadmill or exercise bike
- a spirometer to measure the lung capacity
- a stopwatch to measure the time taken for exercise and the pulse

6. Risks and management of risks

1 A

7. Hazardous substances and control measures

1 Hydrochloric acid is corrosive, which means it attacks and destroys living tissue. So acid will damage the skin if it comes into contact with it.

8. Protective clothing

1 Lab coats protect a person's skin and clothes from splashes, spills and flames. They can also be removed very quickly if needed.
2 Gloves are worn to protect the hands from harmful substances or to prevent the spread of infection if bacterial cultures are being grown. Gloves can also be worn to protect the hands from heat or cold.
3 An irritant is not corrosive but will cause irritation to the skin, eyes or inside the body. So a scientist should wear gloves, goggles and a lab coat when handling these substances.

9. Handling microorganisms

1 He should wear gloves when carrying out this experiment and also sterilise all equipment after it has been used.

He should wear gloves to prevent any bacteria from coming into contact with his skin as he may have a cut on his hand through which the bacteria could enter his blood stream and cause an infection.
He should sterilise all the equipment after use to make sure the bacteria are destroyed so that if another person uses the equipment then they will not be at risk of infection.

10. Microorganisms and wildlife

1 Soil could contain microorganisms. An antibacterial soap will destroy these microorganisms so eliminating the risk of infectious disease.
2 An autoclave heats equipment to high temperatures to kill microorganisms and is a more effective method than using disinfectant.

11. Dependent and independent variables

1 (a) the person's vital lung capacity
(b) the quantity of cigarettes smoked
2 (a) the sunflower height
(b) the quantity of fertiliser used

12. Control variables

1 Measure the volume of hydrochloric acid accurately using a measuring cylinder with appropriate scale markings.
2 The height / angle of the ramp; the type of surface; the point / position of release.

13. Measurements

1 (a) 91 mm
(b) 7 mm
2 0.000 000 01 m

14. Units of measurement

1 B
2 W
3 hertz
4 power

15. Accurate and precise measurements 1

1 Accurate measurements are close to the correct value.
2 They are not precise as the measurements are not close together, there is a difference of between 5 and 10 W between the measurements.

16. Accurate and precise measurements 2

1 The meniscus in a measuring cylinder curves so if you take a reading from the top of the meniscus it will be higher than the reading from the bottom of the meniscus, and not accurate.

17. Range and number of measurements

1 She could state the range of tests she will carry out; such as add 100 g each time until she has 6 different extensions, as this will give a large enough range of values to show a pattern on the graph.
She could take more than one measurement for each weight that is hung on the spring to check that these results are reliable.

18. Writing a method

1 No, this method does not test the hypothesis as the method doesn't involve weighing the trolley, so from this investigation it is not possible to find out how weight affects the speed of an object.

19. Hypotheses

1 Although the hypothesis relates the independent variable (concentration of penicillin) to the dependent variable (growth in bacteria) the science behind it is not accurate as lower concentrations of penicillin will inhibit the growth of bacteria less.

20. Qualitative and quantitative hypotheses

1 (a) quantitative
   (b) qualitative

2 The more fertiliser that is added to soil, the higher the sunflowers will grow.

21. Planning an investigation

1 The longer the piece of wire the greater the resistance it will create in a circuit. This is because in a longer piece of wire the current will have a greater distance to travel, which creates more resistance.

The equipment needed includes:
- an ammeter
- a voltmeter
- 5 wires of different lengths
- 2 crocodile clips
- a battery (1.5 volts)
- a metre ruler.

I will use 5 different lengths of wire, 5 mm, 10 mm, 15 mm, 20 mm, 25 mm. I will measure the current and voltage through the circuit when each wire is attached. I will need to make sure I do not touch the wire when the cell is turned on.

The variables that need to be controlled are that the same circuit and battery are used, and the wire is made from the same material for each length of wire.

22. Learning aim A sample questions

This is an example answer. It is not the only the correct answer.
The more salt added to the water, the lower the boiling point.
Equipment: table salt, measuring cylinder, distilled water, Bunsen burner, tripod, gauze, beaker, thermometer, stirring rod, spatula, digital balance.

Method:

1 Measure out 100 cm³ of distilled water in a measuring cylinder.
2 Pour the water into a beaker.
3 Place the beaker on top of some gauze on a tripod over a Bunsen burner.
4 Heat the water in the beaker and measure the temperature when the water boils.
5 Measure out 5g of table salt using a spatula and digital balance.
6 Add the measured salt to the boiling water and stir.
7 Measure the temperature of the boiling water with the salt in it. Record the highest temperature reading.
8 Repeat for 10 g, 15 g , 20 g and 25 g of salt.

Learning aim B

23. Tables of data

1 The higher the temperature, the greater the mass of sugar.

24. Identifying anomalous results from tables

1 (a) The result at 15 g mass of salt is an anomaly.
   (b) The boiling point decreases as the quantity of salt is increased. However at 15 g the boiling point is only 1 °C less than for 10 g, where as the differences between the other boiling points is at least 4 °C.

25. Identifying anomalous results from graphs

1 He may have fallen over in the race, which would have slowed him down a lot.

26. Excluding anomalous results

1 (a) The reading 12.2 cm for person C is anomalous.
   (b)

Person	Jump height (cm)	Mean jump height (cm)
A	20.2, 22.4, 21.2	21.3
B	15.6, 16.8, 17.9	16.8
C	12.2, 22.5, 21.9	22.2
D	18.9, 19.8, 17.9	18.9

The anomaly should not be included in the mean calculation as it will give an inaccurate mean value.

27. Calculations from tabulated data

1

Concentration of HCl (M)	Mean temperature change (°C)
1	4.33
2	6.33
3	7.67
4	11.0
5	13.5

28 Calculations from tabulated data – using equations

1 (a) 10.4°C
   (b) $50 \times 4.2 \times 10.4 = 2184$ J

29. Significant figures

1 1.67 ohms
2 12 volts

30. Bar charts 1

1 (a) bar chart
   (b) line graph
   (c) line graph

31. Bar charts 2

1 Each axis should be labelled and also have units. The chart should have a title.

32. Pie charts

1 Pie charts are good for displaying data for around 6 or fewer categories. There are more than 6 categories in this data set so a bar chart would be a clearer way of presenting this information.

33. Line graphs 1

1

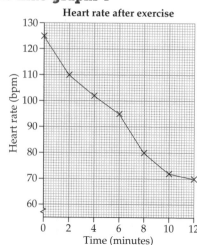

Heart rate after exercise

68

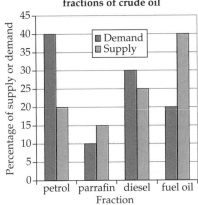

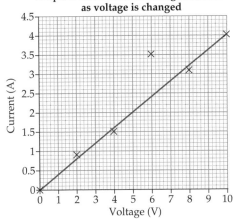

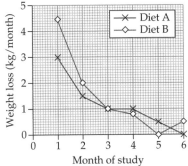

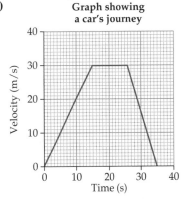

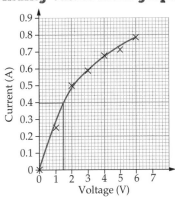

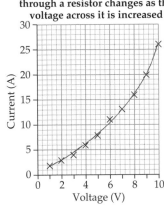

34. Line graphs 2

1

Weight loss for different diets

(Graph: x-axis "Month of study" 0–6, y-axis "Weight loss (kg/month)" 0–5, with two series: Diet A (×) and Diet B (◇))

35. Straight lines of best fit

1

Graph to show current through a resistor as voltage is changed

(Graph: x-axis "Voltage (V)" 0–10, y-axis "Current (A)" 0–4.5)

36. Curves of best fit

1 (a)

A graph to show how the current through a resistor changes as the voltage across it is increased

(Graph: x-axis "Voltage (V)" 0–10, y-axis "Current (A)" 0–30)

(b) As only two or three of the points seem to lie close to a straight line, a curve of best fit shows the pattern of the results better.

37. Selecting types of graph 1

1

Demand and supply in fractions of crude oil

(Bar chart: x-axis "Fraction" with categories petrol, parrafin, diesel, fuel oil; y-axis "Percentage of supply or demand" 0–45; series Demand and Supply)

2 The independent variable is qualitative and a bar chart is good for comparing values.

38. Selecting types of graph 2

1 A bar chart or line graph would be better as pie charts should be used for 6 or fewer data values in a set. In this chart there are 11 values so it is difficult to see what information is being presented.

39. Finding values from graphs 1

1 (a)

(Graph: x-axis "Voltage (V)" 0–7, y-axis "Current (A)" 0–0.9, with curve of best fit)

Estimated current = 0.4 A.

(b) The curve of best fit could be extended to predict results that are beyond the data set included in this investigation to estimate the current at 8 V.

40. Finding values from graphs 2

1 (a) 68 cm

(b) 80 cm

41. Calculating gradients from graphs

1 Acceleration = $\dfrac{15}{6}$ = 2.5 m/s^2

42. Calculating areas from graphs

1 (a)

Graph showing a car's journey

(Graph: x-axis "Time (s)" 0–40, y-axis "Velocity (m/s)" 0–40)

(b) 150 m

43. Why anomalous results occur

1 He would not be measuring the quantities accurately as the scales are not set to zero, so he will not be finding out the true weights of the objects which creates anomalous results.

44. Positive correlation

1 There is a positive correlation as when the independent variable (length) increases, the dependent variable (mass) increases and the line slopes upwards.

ANSWERS

45. Negative correlation

1 There is a negative correlation as the independent variable (number of cigarettes smoked) increases, the dependent variable (the volume of oxygen taken up) decreases and the line slopes downwards.

46. Direct proportion

1 (a) When two variables are directly proportional it means that when one changes the other changes in the same way, by the same proportion or factor.

(b)
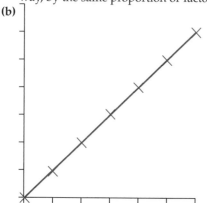

47. Inverse proportion

1 The variables are inversely proportional to another, one variable decreases as the other variable increases.

48. Using evidence to draw conclusions 1

1 The temperature of the cold pack increases over time. However, as the graph curves upwards the temperature change is not directly proportional to time.

49. Using evidence to draw conclusions 2

1 The greater the volume of sodium hydroxide added to the acid, the higher the temperature change. It can be concluded that the more sodium hydroxide is added then the greater the exothermic reaction, which results in the higher temperature change.

50. Learning aim B sample questions

1
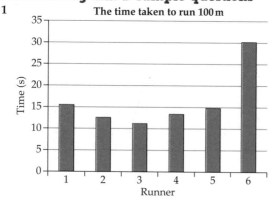

Learning aim C

51. Writing a conclusion 1

1 1 The hypothesis.
   2 Whether the hypothesis is accepted or rejected.
   3 Results or data to confirm why the hypothesis has been rejected or accepted.
   4 Any relationship between the variables.
   5 Scientific theory to justify the results.
   6 A concluding sentence to sum up the results of the investigation.

52. Writing a conclusion 2

1 The hypothesis was 'the greater the force was applied to the trolley the faster the acceleration'. The hypothesis can be accepted as the results show that the greater the force the greater the acceleration. The greatest acceleration is $3.82\,\text{m/s}^2$ which came from the applied force of $0.6\,\text{N}$ and the lowest acceleration is $1.26\,\text{m/s}^2$ which came from the lowest force of $0.2\,\text{N}$.

The variables show that there is a positive relationship between acceleration and force, so the acceleration of an object rises as more force is applied.

53. Inferences

1 The greater the depth of the sea, the lower the number of different species of fish there will be.
2 A

54. Support for a hypothesis 1

1 No, the hypothesis cannot be supported as the distance travelled increased only up to the slope angle of 50 degrees. When the angle was greater than 50 degrees, the distance travelled decreased.

55. Support for a hypothesis 2

1 (a)
Force and acceleration of a trolley travelling down a ramp

(b) There is a positive correlation between force and acceleration.

(c) The investigation shows that the acceleration of an object depends on the size of the force applied to it. If the force doubles, then the acceleration will double. This relationship is described by the following equation: force = mass × acceleration.

56. Support for a hypothesis 3

1 D

57. Strength of evidence

1 (a) The data points on a graph would follow or be close to a line of best fit.
   (b) The data points would follow a trend but would not all be very close to the line of best fit.
   (c) The data points would not follow any trend so a line of best fit could not be drawn.

58. Sufficient data 1

1 In this investigation each person only carries out the test once. An investigation should include repeating measurements at least three times to check that the results are repeatable. The mean value can then be found as this is more accurate than a single value.

59. Sufficient data 2

1 The data collected are subjective observations of an individual's feelings, which means that the data are not reliable. This is because the strength of people's feelings will differ between different people – what feels sad to one person may be very happy to another.

60. Validity of a conclusion 1

1 This is not a valid conclusion as the difference in height between the two plants is only 0.5 cm. This is not enough to confirm that the difference in light has a significant effect on growth. Also, only two plants were used so there is not enough data to draw a valid conclusion.

61. Evaluating an investigation 1

1 He could have measured out the mass of ethanol incorrectly or, he could have read the thermometer incorrectly to produce this anomalous result.

62. Evaluating an investigation 2

1 Only one set of data has been obtained for each value. This means the data may not be reliable.

63. Improving an investigation 1

1 It is a good idea to repeat the measurement a number of times for the same set of variables and calculate a mean. By working out a mean the data will be closer to the true value.

64. Improving an investigation 2

1 She could increase the sample size by increasing the number of independent values for the mass of salt by extending the range. If there are only three data points then it is very difficult to find a trend in the data so the conclusion may not be valid. Extending the range will help to find out if as more salt is added, the boiling point of water continues to decrease. This will give her a greater chance of getting results that helps her to draw a conclusion to prove or disprove the hypothesis.

65. Extending an investigation

1 She could change the conditions of the investigation by finding out how heating one of the selected balls affects it's the rebound height.

66. Learning aim C: 6-mark questions

This is an example answer. It is not the only correct answer.

1 The internal pressure of each ball could be tested when the ball is heated to different temperatures. This will help to find out if there is a change in internal pressure in the squash ball that correlates to the higher rebound. If this was the case, a conclusion could be drawn that this was the cause of the change in rebound height.

Your own notes

Your own notes

Your own notes